THE PROBLEM OF

Dedicated to the memory of
Wesley Salmon,
who died tragically on 22 April 2001,
and who made a lasting contribution to the topics of this volume

The Problem of Realism

Edited by

MICHELE MARSONET
University of Genoa, Italy

LONDON AND NEW YORK

First published 2002 by Ashgate Publishing

Reissued 2018 by Routledge
2 Park Square, Milton Park, Abingdon, Oxon OX14 4RN
711 Third Avenue, New York, NY 10017, USA

Routledge is an imprint of the Taylor & Francis Group, an informa business

Publisher's Note
The publisher has gone to great lengths to ensure the quality of this reprint but points out that some imperfections in the original copies may be apparent.

Disclaimer
The publisher has made every effort to trace copyright holders and welcomes correspondence from those they have been unable to contact.

A Library of Congress record exists under LC control number: 2001022840

ISBN 13: 978-1-138-73728-0 (hbk)
ISBN 13: 978-1-138-73718-1 (pbk)
ISBN 13: 978-1-315-18550-7 (ebk)

Contents

List of Contributors

Evandro Agazzi is currently Professor of Philosophy at the University of Genoa, after having been Professor also at the University of Fribourg (Switzerland) and Visiting Professor at other universities in Europe and America. He is President of the International Academy of Philosophy of Science (Brussels), was President (and is now Honorary President) of the International Federation of the Philosophical Societies, and of the International Institute of Philosophy (Paris). He has published in several languages, as author or editor, more than fifty books, and more than 600 papers in books and journals. Among his recently edited books, *Realism and Quantum Physics* (1997) and *The Reality of the Unobservable* (2000) are close to the problems treated in the present volume.

Pierre Cruse took his undergraduate degree in physics and philosophy at the University of Glasgow. He recently completed his PhD dissertation, *Describing Science: The Descriptivist Approach to Scientific Realism*, at King's College, University of London, under the supervision of David Papineau and Gabriel Segal. He is currently a Teaching Fellow at King's College, University of London.

Michael Devitt did a BA at the University of Sydney and then went to Harvard University to do a PhD. He taught at the University of Sydney from 1971 until 1987 when he was appointed a Professor of Philosophy at the University of Maryland. In 1999 he moved to the City University of New York taking the position of Executive Officer of the Philosophy Program at the Graduate Center, where he was promoted to Distinguished Professor. He has written extensively on realism and the philosophy of language and mind, including the following books: *Designation* (1981), *Realism and Truth* (2nd ed. with new Afterword 1997), *Coming to Our Senses: A Naturalistic Defense of Semantic Localism* (1996), and (with Kim Sterelny) *Language and Reality: An Introduction to the Philosophy of Language* (2nd ed. 1999).

Christopher Hughes (BA Wesleyan, PhD Pittsburgh) has taught at King's College, University of London since 1984; he had previously taught at Cornell University. He is especially interested in metaphysics, mediaeval philosophy, and the philosophy of religion. He has written a book on Aquinas' philosophical theology (*On A Complex Theory of a Simple God*, 1989), and numerous articles in the areas mentioned above. His present interests include various topics concerning the persistence and identity of animate and inanimate objects, mediaeval and contemporary attempts to reconcile God's omniscience with the openness of the future, and the compatibility or otherwise of determinism with free will.

Michele Marsonet received his degrees from the Universities of Genoa and Pittsburgh, and is currently Professor of Philosophy at the University of Genoa. His interests include metaphysics, philosophy of science, philosophy of logic, and political philosophy. He has been Visiting Professor in European, American, and Australian universities, and is President of the Ligurian Philosophical Association. He has published books and articles in several languages. His latest books in English are *Science, Reality, and Language* (1995) and *The Primacy of Practical Reason* (1996).

David Papineau is Professor of Philosophy of Science at King's College, University of London. He has previously held posts at Cambridge, Reading, and Macquarie Universities. From 1993 to 1995 he was President of the British Society for the Philosophy of Science, and from 1994 to 1998 he was Editor of the *British Journal for the Philosophy of Science*. He has written widely in the philosophy of science and the philosophy of psychology. He is the author of *For Science in the Social Sciences* (1978), *Theory and Meaning* (1979), *Reality and Representation* (1987), *Philosophical Naturalism* (1993), and *Introducing Consciousness* (2000), and he edited *Philosophy of Science* (1996). His new book, *Thinking about Consciousness*, will be published in 2001.

Paolo Parrini is Professor of Philosophy at the University of Florence and has been President of the Italian Society of Analytic Philosophy. His research is devoted to contemporary analytic philosophy, Kant's and Husserl's epistemological thought, and the various aspects of nineteenth- and twentieth-century scientific and epistemological thought. His most recent book is *Knowledge and Reality. An Essay in Positive Philosophy* (1998). He edited *Kant and Contemporary Epistemology* (1994) and is currently editing, with Wesley C. Salmon, a volume of essays on logical empiricism.

Mark Sainsbury is Susan Stebbing Professor of Philosophy at King's College, University of London and a Fellow of the British Academy. Previous appointments were at Bedford College, University of London, and at the University of Essex. His publications lie mostly within the areas of philosophy of language and philosophical logic, and his most recent books are *Paradoxes* (2nd ed. 1995) and *Logical Forms* (2nd ed. 2000). He is currently writing a book entitled *Reference without Referents*, and has research leave made possible by the Leverhulme Trust. He was Editor of *Mind* for ten years until 2000.

Wesley C. Salmon was University Professor of Philosophy, Emeritus, at the University of Pittsburgh. Having received his PhD from the University of California at Los Angeles in 1950, he devoted his main research efforts to causation/explanation, probability/induction/confirmation, and space-time. Among his major works are *Scientific Explanation and the Causal Structure of the World* (1984) and *Causality and Explanation* (1998). He was former President of the American Philosophical Association (Pacific Division), the Philosophy of Science Association, and the International Union of History and Philosophy of Science.

Howard Sankey studied philosophy at the University of Otago, before completing his PhD at the University of Melbourne, where he is currently Associate Professor in the Department of History and Philosophy of Science. He is the author of numerous articles on topics such as incommensurability, scientific realism, relativism, and the rationality of scientific theory choice, as well as two books on these issues: *The Incommensurability Thesis* (1994) and *Rationality, Relativism and Incommensurability* (1997). He is the editor of *Causation and Laws of Nature*, and co-editor (with Robert Nola) of *After Popper, Kuhn and Feyerabend: Recent Issues in Theories of Scientific Method*, and (with Paul Hoyningen-Huene) of *Incommensurability and Related Matters*.

Chapter One

Introduction

Michele Marsonet

This collection of essays deals with the main problems involved in the realism–anti-realism debate, and I would like to stress that the central topic is realism 'as such', which explains why issues connected to both metaphysical and scientific realism play an important role in the contributions to the volume. This, of course, assuming that scientific realism can be clearly distinguished from metaphysical realism in Hilary Putnam's sense of the term,[1] a thesis that some, nowadays, do not feel inclined to endorse. The following are a few essential points emerging from the current debate on the problem of realism.

It is often stated that forming a 'conception of reality' is something dependent upon language. If a subject is to have a view about reality, he must have access to an inter-subjective standard provided by a social-linguistic world.[2] It is only in learning a language that one gains the ability to respond conceptually to the world, because only then can a person have responses assessed by social norms.

It follows – some philosophers add – that our conception of reality depends upon factors that are not totally describable by science. That is to say, we should accept the fact that there can be no completely neuro-computational[3] or mechanical account of how we come to have such a conception of the real. This seems to jeopardize attempts to base realism on naturalism. As we know, many reject this kind of story, and insist instead on the possibility of a straight naturalistic approach.[4] The debate on this issue is as open as ever, as the chapter by Michael Devitt included in the collection clearly shows.

Another important point to be made about the nature of realism as such is that, according to some authors, what basically differentiates it is the epistemological thesis that the entities to which ontological commitment is made (by human beings) exist independently of any knowledge of them.[5] But then it looks possible – some even say easy – to turn the metaphysical thesis related to the existence of such entities, into an epistemological one. If so, what is at stake in the realism–anti-realism debate is neither a question of metaphysics nor of semantics, but of epistemology.[6] Michael Dummett, instead, believes that the true nature of metaphysical disputes about realism is that 'they are disputes about the kind of meaning to be attached to various types of sentences'.[7] Furthermore, we all know that the 'independence thesis' plays a key role in metaphysical realism.[8] As Hilary Putnam remarks:

> On this perspective, the world consists of some fixed totality of mind-independent objects. There is exactly one true and complete description of 'the way the world is'. Truth involves some sort of correspondence relation between words or thought-signs and external things and sets of things. I shall call this perspective the *externalist* perspective, because its favourite point of view is a God's Eye point of view.[9]

However, the use of the term 'independence' implies 'independence from something', most notably the mind. Thus, according to this trend of thought, it looks as if we cannot avoid reference to minds even when formulating the most basic tenet of metaphysical realism.

The third point to be raised is the following. It is usually held that a very important issue in the general problem of realism – and in the realism–anti-realism debate as well – is anti-realism about the physical world. But to what extent is this assertion correct? Sometimes philosophers charged with being anti-realists or idealists turn out not to be clearly so if one reads their works carefully. Think of the alleged 'linguistic idealism' in Wittgenstein's thought:[10] from several points of view, it is questionable to interpret Wittgenstein in this way.

The fact is that it seems wrong to equate anti-realism and idealism.[11] It is one thing to claim that entities are made up of mental items, as some classical idealists do; quite another to say that our access to reality is always mediated by epistemic and mind-involving constraints.[12] What distinguishes such a view from a realist one is that, unlike the realist, the anti-realist can make no sense of metaphysical claims without resorting to some kind of supporting epistemology. The important point, in sum, is that to reject metaphysical realism (at least in Putnam's sense of the term, as specified above) is not the same as endorsing the view that there are no mind-independent objects in the world.

The fourth point is that it may turn out to be difficult to be realists about both common-sense objects and scientific entities.[13] According to some authors they belong, in fact, to two different conceptual schemes (as Sellars, for example, claims[14] and van Fraassen denies).[15] But it is true that any attempt at reconciling the two schemes, or at reducing them to one, gives rise to problems which admit of no easy solution. It has often been claimed, in fact, that the ontology of the two schemes seems to be incompatible in many respects, and that one scheme (usually common-sense) is fated to be replaced by the other (science).[16] A scientifically oriented philosopher might at this point be tempted to affirm the absolute superiority of the scientific world-view, but there are many doubts about the possibility of attributing to science such a primary role in assessing any kind of conceptual scheme.

Strictly connected to these remarks is the issue of the relations between metaphysics and epistemology. Many realists claim that straightforwardly metaphysical questions should be kept separate from epistemological issues. But can we really do this? And what does the expression 'straightforwardly metaphysical' mean? After all, any kind of ontology is characterized by the fact that the things of nature are seen by us in terms of a conceptual apparatus that is always – and substantially – influenced by elements referring to minds. It might be stated that the task of metaphysics is to discover what kinds of entities make up the world, while epistemology's job is to find out what are the principles by which we get to know reality. It is obvious, however, that if the conceptual apparatus is at work in any case, our access to reality always entails the involvement of the mind. Naturalization of the mind – and of its activities – is an obvious answer to this problem[17] but, as I said at the beginning, agreement on naturalization cannot be taken for granted.

We must thus face the question of 'ontological pluralism', a basic tenet of contemporary neopragmatism.[18] Ontological pluralism is in turn connected to the existence of alternative ways of conceptualizing the world. It has, in fact, been noted that our world-view is not neutral, and that beings whose experiential modes are (substantially) different from ours must conceptualize reality differently. If so, it would seem to follow that any objective ontology should be left in the background.

The last point is the following. We are confronted, eventually, by a crucial question: what kind of realism – if any – are we allowed to endorse? It is often stated that, in order to provide realism with a solid foundation, we need to have recourse to a reality that is totally independent of thought (let alone of language). This is taken to be the key thesis of realism. But many philosophers reply that, even when we imagine a world totally devoid of human presence, we are bound to use concepts. From this point of view, conceptualization does not seem to be an optional, but a structural component of our nature.

It must be noted that such remarks do not entail anti-realism. The opinion that, due to our cognitive position in the world and its limitations, the perspective provided by the conceptual framework we employ cannot be transcended, is rather widespread today. This amounts to saying that, although the world does not need our participation in order to be, our epistemic access to the world is given by such participation. Any description, thus, is bound to be determined by our operational perspectives.

Nothing prevents us from claiming that objective reality – a reality which does not depend on our cognitive capacities – is there. But, of course, a strong realist is not likely to be satisfied with such an answer, because this position corresponds, roughly, to what Devitt defines as 'weak, or fig-leaf, realism', that is to say, a commitment to there being just 'something' independent of us.[19] A strong realist cannot accept this solution, but the question to be asked is: are we in a position to say anything more? A commitment to there just being 'something' independent of us is enough to establish, at least, the basic tenet of metaphysical realism in Putnam's sense. On the other hand, to say more than this means to get involved in disputes which stem not only from the philosophical field, but also from science, quantum theory being a paradigmatic example.

Let us now briefly overview the content of the various contributions to the volume. Michael Devitt, in 'A Naturalistic Defence of Realism', notes that anti-realism about the physical world is an occupational hazard of philosophy. Most of the great philosophers have been anti-realists in one way or another. Many of the cleverest contemporary philosophers are also: Michael Dummett, Nelson Goodman, Hilary Putnam, and Bas van Fraassen. Yet anti-realism is enormously implausible on the face of it.

The defence of realism depends on distinguishing it from other doctrines and on choosing the right place to start the argument. And, according to the author, the defence of that choice depends on naturalism. In the first section of his paper Devitt says what realism is, distinguishing it from semantic doctrines with which it is often confused. In the second section he considers the arguments for and against realism about observables. In the third section the author considers the arguments for and against realism about unobservables, namely 'scientific realism'.

'Metaphysical and Scientific Realism' is the title of Evandro Agazzi's chapter. He starts by noting that the expression 'metaphysical realism' is known for having been used especially by Putnam, who means by it the philosophical perspective according to which 'the world consists of some fixed totality of mind-independent objects. There is exactly one true and complete description of the way the world is'. Truth involves some sort of correspondence relation between words or thought-signs and external things and sets of things.

In contrast with this view Agazzi proposes to call 'metaphysical realism' the position that does not subscribe to the two presuppositions that we directly know only our representations, and that there is a 'reality' behind these representations. He claims that no evidence nor argument have ever been proposed for such claims, and what we can, on the contrary, maintain is that reality is 'present' to our senses and thinking. To say that what is present is not reality is a claim whose burden falls on those who put forth the claim, and it is not easy to understand how they could substantiate it. All this does not entail that 'the whole of reality' is present in any act of knowledge. On the contrary, only certain aspects or attributes of reality are present in our acts of knowledge.

Since human beings are endowed with cognitive capabilities that overstep those of other living beings, it is obvious that certain features or attributes of reality can be present to our intellect, and these are those features that we call 'universal' or 'abstract'. If we call 'intuition' the immediate presence of something to our cognitive capacities, we must conclude that, besides the sensorial intuition, we are endowed also with an intellectual intuition.

The characterization of realism defended by Agazzi can be summarized as the thesis of the 'knowability and intelligibility of reality'. This realism is external in Putnam's sense, but does not coincide at all with his 'God's Eye view'. While admitting that what is 'immediately known' is real, we also concede that the domain of reality is broader than the domain of what is immediately present. At the same time, we recognize that the intellect can lead us to determine features of reality that are not immediately present. Agazzi's claim is that metaphysical realism is necessary for the construction of metaphysics in a disciplinary sense, but is a kind of methodological prerequisite for the construction of science as well and, in particular, justifies a scientific realism that is at the same time 'internal' and 'external'.

Howard Sankey, in 'Realism, Method, and Truth', takes into account the relation between scientific method and truth. In other words, are we justified in accepting a theory that satisfies the rules of scientific method as true? Such questions divide realism from anti-realism in the philosophy of science.

Scientific realists take the methods of science to promote the realist aim of correspondence truth. Anti-realists either claim that the methods of science promote lesser epistemic goals than realist truth, or else they reject the realist conception of truth altogether. In his chapter, Sankey proposes a realist theory of the relation between scientific method and truth. The theory consists of three basic elements: (1) a naturalistic treatment of epistemic normativity, (2) an instrumentalist conception of the nature of methodological rules, and (3) an abductive argument for the truth-conduciveness of the rules of scientific method.

'The Real Distinction between Persons and their Bodies' is the theme dealt with by Christopher Hughes. Are my body and I two different things, or one and the same thing? Philosophers have offered a variety of arguments to the effect that they are two different things. One is the argument from temporal discernibility, according to which I and my body are distinct because we do not exist at all the same times (it will outlast me). Another is the argument from modal discernibility, according to which I and my body are distinct because we do not exist in all the same possible worlds.

The argument from modal discernibility is in turn found in two versions – a Cartesian version, and a Lockean one. (The Cartesian version, but not the Lockean, turns on the possibility of disembodied existence.) Although some philosophers (for example Bernard Williams, Michael Ayers, and Fred Feldman) have resisted the arguments from temporal and modal discernibility offered by Kripke and others, there is something of a philosophical consensus that (at least some of) those arguments are successful in showing the distinctness of me from my body. Hughes argues that Williams and Feldman fail to successfully defend the identification of persons with their bodies from the argument from temporal discernibility.

But, as he tries to show, there is a different and defensible way of blocking that argument. Moreover, he argues, Cartesian arguments from modal discernibility fail to show that I and my body are distinct. The strongest argument for the distinctness of me from my body is the Lockean argument from modal discernibility; but it, too, Hughes argues, falls short of cogency. He concludes that the identification of persons with their bodies remains philosophically defensible.

In 'A Realistic Account of Causation', Wesley Salmon notes that David Hume's classic account of causation places cause–effect relations in the human imagination. His chapter aims to show that causal relations are not merely a product of human thought; they exist in the physical world and enjoy the same sort of reality as tables, rocks, and cats. After critically reviewing various unsuccessful post-Humean attempts to restore the objective status of causation, including J. L. Mackie's famous account in terms of INUS conditions, Salmon temporarily sets aside such 'event' terms as 'cause' and 'effect' in order to introduce a process-based analysis.

'Causal interaction' is the fundamental causal concept; it is explicated via such non-causal concepts as spatio-temporal intersection, process (causal or pseudo-), and qualitative difference. With the aid of interaction, the concept of causal transmission is introduced. This, in turn, allows us to introduce the concept of causal (as opposed to pseudo-) process. Salmon maintains that many causal processes are open to ordinary observation, and offers empirical tests to distinguish causal from pseudo-processes. He claims that causal processes are precisely the causal connections Hume sought in vain. Finally, 'cause' and 'effect' are reintroduced on the basis of the foregoing causal concepts.

The main aim of Paolo Parrini's contribution, 'Realism and Anti-Realism from an Epistemological Point of View', is to discuss the contrast between realism and anti-realism from an epistemological point of view. Anti-realists have been able to give good reasons for the following three main theses: (1) our knowledge cannot do without historically variable subjective epistemic assumptions; (2) what we know as 'object' is in fact the result of a synthesis of experience effected on the basis of different kinds of epistemic assumptions; (3) there is an apparently indissoluble tie between scepticism and realism.

To the above-mentioned three main anti-realist theses correspond three alternative theses put forward by the supporters of realism: (1) the notion of 'object of knowledge' – even if it is meant in an absolute, metaphysical sense – is neither meaningless nor internally incoherent; (2) the structure and the development of knowledge depend on a 'given' which is not completely determinable by the knowing subject; (3) anti-realists have never produced good arguments in order to establish that our epistemic conditions of knowledge could not as well be structural properties of objects 'in themselves'.

This kind of balance of the arguments seems to suggest that the contrast between realism and anti-realism cannot be solved by appealing to logical-analytical arguments and to proofs of a factual kind. We can try to give an answer to the question only by resorting to a theoretical reconstruction of a 'non-coercive' kind that aims at competing with rival proposals hoping to be judged as the best one, at least on the whole. The last part of the chapter describes a proposed solution based on the idea of the regulative and empty character of the notions of objectivity and truth, and on the acceptance of empirical realism and the refusal of metaphysical realism.

In 'Realism vs Nominalism about the Dispositional–Non-Dispositional Distinction', Mark Sainsbury observes that one line of thought that Hume held in attacking the idea of power may have been this: a power would

have to be both intrinsic (in or 'on' the object which possessed it), in order
to do its causal and explanatory work; yet, at the same time it seems it must
be non-intrinsic (it is not 'compleat in itself', and 'points out' to something
else). Hence the very idea of power is implicitly contradictory.

Dispositional nominalism is the view that the distinction between
dispositional and non-dispositional is a distinction among predicates but
does not distinguish properties with different natures. Dispositional
realism is the more conventional view that a dispositional property is
different in kind from a non-dispositional one. Dispositional nominalism
would solve Hume's problem; but would it have other merits?

Sainsbury argues that dispositional nominalism and dispositional
realism are on a par with respect to giving a proper account of the causal role
of properties, but that dispositional nominalism can resolve a problem
which realism cannot: for the realist, it seems that the fundamental nature of
the world must involve non-dispositional properties, yet fundamental
explanations in physics appear to appeal to dispositional properties. On the
nominalist view of the distinction, this marks a difference in our vocabulary,
but not a difference at the level of the nature of the properties themselves.

In their chapter 'Scientific Realism without Reference', Pierre Cruse
and David Papineau start from Larry Laudan's claim that scientific
theories are normally not even approximately true, since their central terms
will characteristically fail to refer, and their central claims will therefore
not even be candidates for truth or falsity. A scientific realist holds,
instead, that our most successful scientific theories are at least
approximately true descriptions of the unobservable facts.

Cruse and Papineau argue, against Laudan, that scientific theories can
be approximately true even if their central claims fail to refer, and the
chapter aims to rebut a familiar history-based objection to scientific
realism. According to this objection, scientific realism requires that past
theoretical terms succeeded in referring, yet history shows that reference
failure is the norm. In response Cruse and Papineau suggest a version of
scientific realism – 'Ramsey-sentence realism' – to which the referential
status of theoretical terms is irrelevant. They also argue that this Ramsey-
sentence realism is equivalent to scientific realism more standardly
construed, though they regard this as a further thesis which is not essential
to Ramsey-sentence realism itself.

Finally, in my own contribution 'The Limits of Realism', I explore to
what extent we are allowed to draw a neat borderline between ontology
and epistemology. A positive answer to this question poses many
problems, as many contemporary thinkers have noted. In particular, our
ontology is characterized by the fact that the things of nature are seen by us

in terms of a conceptual apparatus that is inevitably influenced by mind-involving elements. The distinction between the natural world on the one hand, and the social-linguistic world on the other is difficult because we began to identify ourselves and the objects that surround us only when the social-linguistic world emerged from the natural one, and this in turn means that our criteria of identification are essentially social and linguistic.

After reviewing such related problems as ontological pluralism, the relation between science and common-sense, and the contrast between strong and weak realism, I argue in favour of a 'minimal' sort of realism. Such a stance, while admitting the presence of an ontologically independent reality, also recognizes the importance of the distinction between reality-as-such and reality-as-known-by-us.

The chapters included in this collection were read as papers at the international conference 'The Problem of Realism', held in Genoa on 29–30 November 1999. They represent a fair variety of positions, thus providing a pluralism which is always valuable in meetings of this kind. Many thanks are due to those who helped me in organizing the conference, and to the Italian National Research Council which kindly agreed to host it. It is important to note, in this regard, that the Centre for Studies on Contemporary Philosophy of the Italian National Research Council sponsored the initiative. Thanks are also due to the Rector of the University of Genoa, along with the Faculty of Humanities and the Department of Philosophy, which all sponsored the conference and generously funded it.

Furthermore, I would like to thank Howard Sankey for revising some of the papers written by Italian authors and, especially, Mark Sainsbury for both his global revision of the text and the precious assistance he gave me in the editorial process. Ginny Sturdy Watkins prepared the camera-ready text and kindly helped me in overcoming all technical problems.

Notes

1 Putnam (1978, 1981 and 1990). For a different meaning of 'metaphysical realism' see E. Agazzi's contribution to this volume (ch. 3).
2 This point has been stressed by Donald Davidson and Wilfrid Sellars in many of their works. See especially Davidson (1991) and Sellars (1997).
3 Such as the one endorsed in Churchland (1988 and 1990).
4 This is the case, for example, with Michael Devitt. See Devitt (1991a and 1991b), and his contribution to this volume (ch. 2).
5 A strong defence of this thesis can be found in Trigg (1989).
6 This is what Anthony Grayling argues in Grayling (1997).
7 Dummett (1991, p. 14).

8   At least in the version of 'metaphysical realism' made popular by Hilary Putnam. See note 1 above.
9   Putnam (1981, p. 49).
10 Bernard Williams argues in favour of Wittgenstein's 'linguistic idealism' in Williams (1974). A recent rebuttal of his thesis is contained in Hutto (1999).
11 As Anthony Grayling notes in Grayling (1997).
12 A clear example of the second position is contained in Rescher (1973).
13 Michael Devitt endorses this position without hesitation in the second chapter of this volume and in his works mentioned in note 4 above.
14 See Sellars (1963 and 1968).
15 van Fraassen (1990).
16 See Feyerabend (1995).
17 In fact, this is the strategy favoured by Michael Devitt in his contribution to this volume.
18 See, for example, Rorty (1982 and 1999).
19 Devitt (1991a).

References

Churchland, P. M. (1988), *Matter and Consciousness: A Contemporary Introduction to the Philosophy of Mind*, Cambridge, MA/London, MIT Press.
Churchland, P. M. (1990), *Scientific Realism and the Plasticity of Mind*, Cambridge, Cambridge University Press.
Davidson, D. (1991), 'Three Varieties of Knowledge', in *A. J. Ayer. Memorial Essays*, A. Phillips Griffiths (ed.), Cambridge, Cambridge University Press, 1991, pp. 153–66.
Devitt, M. (1991a), *Realism and Truth*, Oxford, Blackwell, 2nd ed.
Devitt, M. (1991b), 'Aberrations of the Realism Debate', *Philosophical Studies*, 61, pp. 43–63.
Dummett, M. (1978), *Truth and Other Enigmas*, Cambridge, MA, Harvard University Press.
Dummett, M. (1991), *The Logical Basis of Metaphysics*, Cambridge, MA, Harvard University Press.
Feyerabend, P. K. (1995), 'Reply to Criticism: Comments on Smart, Sellars and Putnam', in P. K. Feyerabend, *Realism, Rationalism and Scientific Method, Philosophical Papers*, vol. 1, Cambridge, Cambridge University Press, 1995, pp. 104–31.
Grayling, A. C. (1997), *An Introduction to Philosophical Logic*, Blackwell, Oxford, 3rd ed.
Hutto, D. D. (1999), *The Presence of Mind*, Amsterdam/Philadelphia, John Benjamins.
Putnam, H. (1978), *Meaning and the Moral Sciences*, London, Routledge and Kegan Paul.
Putnam, H. (1981), *Reason, Truth and History*, Cambridge, Cambridge University Press.
Putnam, H. (1983), 'Why There Isn't A Ready-Made World', in H. Putnam, *Realism and Reason, Philosophical Papers*, vol. 3, Cambridge, Cambridge University Press, 1983, pp. 205–28.
Putnam, H. (1991), *The Many Faces of Realism*, La Salle, Illinois, Open Court, 3rd ed.
Rescher, N. (1973), *Conceptual Idealism*, Oxford, Blackwell.
Rorty, R. (1982), *Consequences of Pragmatism*, Minneapolis, University of Minnesota Press.
Rorty, R. (1999), *Philosophy and Social Hope*, London, Penguin.

Sellars, W. (1963), 'Philosophy and the Scientific Image of Man', in *Science, Perception and Reality*, London, Routledge and Kegan Paul, 1963, pp. 1–40.

Sellars, W. (1968), *Science and Metaphysics: Variations on Kantian Themes*, London, Routledge and Kegan Paul.

Sellars, W. (1997), *Empiricism and the Philosophy of Mind*, Cambridge, MA/London, Harvard University Press.

Trigg, R. (1989), *Reality at Risk: A Defence of Realism in Philosophy and the Sciences*, New York/London, Harvester Wheatsheaf, 2nd ed.

van Fraassen, B. (1990), *The Scientific Image*, Oxford, Clarendon Press, 4th ed.

Williams, B. (1986), 'Wittgenstein and Idealism', in *Understanding Wittgenstein*, London, Macmillan, 1974, pp. 76–95.

Chapter Two

A Naturalistic Defence of Realism

Michael Devitt

Anti-realism about the physical world is an occupational hazard of philosophy. Most of the great philosophers have been anti-realists in one way or another. Many of the cleverest contemporary philosophers are also: Michael Dummett, Nelson Goodman, Hilary Putnam, and Bas van Fraassen. Yet anti-realism is enormously implausible on the face of it.

The defence of realism depends on distinguishing it from other doctrines and on choosing the right place to start the argument. And the defence of that choice depends on naturalism. In the first section I shall say what realism is, distinguishing it from semantic doctrines with which it is often confused. In the next section I shall consider the arguments for and against realism about observables. In the final section I shall consider the arguments for and against realism about unobservables, 'scientific' realism. The discussion is based on my book, *Realism and Truth* (1997; unidentified references are to this work).

What is realism?

A striking aspect of the contemporary realism debate is that it contains almost as many doctrines under the name 'realism' as it contains participants.[1] However, some common features can be discerned in this chaos. First, nearly all the doctrines are, or seem to be, partly semantic. Consider, for example, Jarrett Leplin's editorial introduction to a collection of papers on scientific realism. He lists ten 'characteristic realist claims' (1984b, pp. 1–2). Nearly all of these are about the truth and reference of theories. Not one is straightforwardly metaphysical.[2] However, second, amongst all the semantic talk, it is usually possible to discern a metaphysical doctrine, a doctrine about what there is and what it is like. Thus 'realism' is now usually taken to refer to some combination of a metaphysical doctrine with a doctrine about truth, particularly with a correspondence doctrine.[3]

The metaphysical doctrine has two dimensions, an existence dimension and an independence dimension (ch. 2 and sec. A.1). The existence dimension commits the realist to the existence of such common-sense entities as stones, trees, and cats, and such scientific entities as electrons, muons, and curved space-time. Typically, idealists, the traditional opponents of realists, have not denied this dimension; or, at least, have not straightforwardly denied it. What they have denied is the independence dimension. According to some idealists, the entities identified by the first dimension are made up of mental items, 'ideas' or 'sense-data', and so are not external to the mind. In recent times another sort of idealist has been much more common. According to these idealists, the entities are not, in a certain respect, 'objective': they depend for their existence and nature on the cognitive activities and capacities of our minds. Realists reject all such mind dependencies. Relations between minds and those entities are limited to familiar causal interactions long noted by the folk: we throw stones, plant trees, kick cats, and so on.

Though the focus of the debate has mostly been on the independence dimension, the existence dimension is important. First, it identifies the entities that are the subject of the dispute over independence. In particular, it distinguishes a realism worth fighting for from what I call 'weak, or fig-leaf, realism' (p. 23): a commitment merely to there being 'something' independent of us. Second, in the discussion of unobservables – the debate about scientific realism – the main controversy has been over existence.

I capture the two dimensions in the following doctrine:

'Realism': Tokens of most common-sense, and scientific, physical types objectively exist independently of the mental.

This doctrine covers both the observable and the unobservable worlds. Some philosophers, like van Fraassen, have adopted a different attitude to these two worlds. So, for the purpose of argument, we can split the doctrine in two: 'common-sense realism' concerned with observables, and 'scientific realism' concerned with unobservables.

In insisting on the objectivity of the world, realists are not saying that it is unknowable. They are saying that it is not constituted by our knowledge, by our epistemic values, by our capacity to refer to it, by the synthesizing power of the mind, nor by our imposition of concepts, theories, or languages; it is not limited by what we can believe or discover. Many worlds lack this sort of objectivity and independence: Kant's 'phenomenal' world; Dummett's verifiable world; the stars made by a Goodman 'version'; the constructed world of Putnam's 'internal realism'; Kuhn's world of theoretical ontologies;[4] the many worlds created by the 'discourses' of structuralists and post-structuralists.

Realism accepts both the ontology of science and common sense, and the folk-epistemological view that this ontology is objective and independent. Science and common sense are not, for the most part, to be 'reinterpreted'. It is not just that our experiences are 'as if' there are cats, there are cats. It is not just that the observable world is 'as if' there are atoms, there are atoms. As Putnam once put it, realism takes science at 'face value' (1978, p. 37).

Realism is the minimal realist doctrine worth fighting for. Once it is established, the battle against anti-realism is won; all that remains are skirmishes. Furthermore, realism provides the place to stand to solve the many other difficult problems that have become entangled with it.

Any semantic doctrine needs to be disentangled from realism (ch. 4 and sec. A.2). In particular, the correspondence theory of truth needs to be disentangled: it is in no way constitutive of realism nor of any similarly metaphysical doctrine.[5]

On the one hand, realism does not entail any theory of truth or meaning at all, as is obvious from our definition. So it does not entail the correspondence theory. On the other hand, the correspondence theory does not entail realism. The correspondence theory claims that a sentence (or thought) is true in virtue of its structure, its relations to reality, usually reference relations, and the nature of reality. This is compatible with absolutely any metaphysics. The theory is often taken to require the objective mind-independent existence of the reality which makes sentences true or false. This addition of realism's independence dimension does, of course, bring us closer to realism. However, the addition seems like a gratuitous intrusion of metaphysics into semantics. And even with the addition, the correspondence theory is still distant from realism, because it is silent on the existence dimension. It tells us what it is for a sentence to be true or false, but it does not tell us which ones are true and so could not tell us which particular entities exist.

Realism is about the nature of reality in general, about what there is and what it is like; it is about the largely inanimate impersonal world. If correspondence truth has a place, it is in our theory of only a small part of that reality: it is in our theory of people and their language.[6]

Not only is realism independent of any doctrine of truth, we do not even need to use 'true' and its cognates to 'state' realism, as our definition shows. This is not to say that there is anything 'wrong' with using 'true' for this purpose. Any predicate worthy of the name 'truth' has a 'disquotational' property captured by the 'equivalence thesis'. The thesis is that appropriate instances of

s is true if and only if *x*

hold, where an appropriate instance is obtained by substituting for '*p*' a sentence which is the same as (or a translation of) the sentence referred to by the term substituted for '*s*'.[7] Because of this disquotational property, we can use 'true' to talk about anything, by referring to sentences. Thus we can talk about the whiteness of snow by saying ' "Snow is white" is true'. And we can redefine the metaphysical doctrine 'realism' as follows:

> Most common-sense, and scientific, physical existence statements are objectively and mind-independently true.

This redefinition does not make realism semantic (else every doctrine could be made semantic); it does not change the subject matter at all. It does not involve commitment to the correspondence theory of truth, nor to any other theory. Indeed, it is compatible with a deflationary view of truth according to which, roughly, the equivalence thesis captures all there is to truth.[8] This inessential redefinition exhausts the involvement of truth in constituting realism.[9]

My view that realism does not involve correspondence truth flies so much in the face of entrenched opinion, and has received so little support, that I shall labour the point. I shall do so by considering a fairly typical contemporary statement of 'scientific realism':

> 'Contemporary realism': Most scientific statements about unobservables are (approximately) correspondence-true.

Why would people believe this? I suggest only because they believed something like the following two doctrines:

> 'Strong scientific realism': Tokens of most unobservable scientific types objectively exist independently of the mental and (approximately) obey the laws of science.

> 'Correspondence truth': Sentences have correspondence-truth conditions.

These two doctrines, together with the equivalence thesis, imply contemporary realism. Yet the two doctrines have almost nothing to do with each other. Contemporary realism is an unfortunate hybrid.

Strong scientific realism is stronger than my minimal doctrine, scientific realism, in requiring that science be mostly right not only about which unobservables exist but also about the properties of those unobservables. But the key point here is that both these doctrines are metaphysical, concerned with the underlying nature of the world in general. To accept strong scientific realism we have to be confident that science is discovering things about the unobservable world. Does the success of science show that we can be confident about this? Is inference to the best explanation appropriate here? Should we take sceptical worries seriously? These are just the sort of epistemological questions that have been, and still largely are, at the centre of the realism debate. Their home is with strong scientific realism not with correspondence truth.

Correspondence truth is a semantic doctrine about the pretensions of one small part of the world to represent the rest. The doctrine is the subject of lively debate in the philosophy of language, the philosophy of mind, and cognitive science. Do we need to ascribe truth conditions to sentences and thoughts to account for their roles in the explanation of behaviour and as guides to reality? Do we need reference to explain truth conditions? Should we prefer a conceptual-role semantics? Or should we, perhaps, near enough eliminate meaning altogether? These are interesting and difficult questions (ch. 6 and secs A.12–15), but they have no immediate bearing on scientific realism.

Semantic questions are not particularly concerned with the language of science. Even less are they particularly concerned with 'theoretical' language 'about unobservables'. Insofar as the questions are concerned with that language, they have no direct relevance to the metaphysical concerns of strong scientific realism. They bear directly on the sciences of language and mind and, via that, on the other human sciences. They do not bear directly on science in general. Many philosophers concerned with semantics and not in any way tainted by anti-realism are dubious of the need for a correspondence notion of truth.[10]

Are there atoms? Are there molecules? If there are, what are they like? How are they related to each other? Strong scientific realism says that we should take science's answers pretty much at face value. So there really are atoms and they really do make up molecules. That is one issue. Another issue altogether is about meaning. Do statements have correspondence-truth conditions? Correspondence truth says that they do. This applies as much to 'Cats make up atoms' as to 'Atoms make up molecules'; indeed it applies as much to 'The Moon is made of green cheese'. Put the first issue together with the second and we get a third: is 'Atoms make up molecules' correspondence-true? My point is that this issue is completely derivative from the other two. It arises only if we are wondering about, first, the meanings of sentences ranging from the scientific to the silly; and about, second, the nature of the unobservable world.

Suppose that we had established that correspondence truth was right for the familiar everyday language. Suppose further that we believed that atoms do make up molecules, and the like. Then, of course, we would conclude that correspondence truth applies to 'Atoms make up molecules', and so on, and so conclude that such sentences are correspondence-true. What possible motive could there be for not concluding this? Scientific theories raise special metaphysical questions, not semantic ones.

Strong scientific realism and correspondence truth have very different subject matters and should be supported by very different evidence. Underlying contemporary realism is a conflation of these two doctrines that has been detrimental to both.

It follows from this discussion that a metaphysical doctrine like realism cannot be attacked simply by arguing against certain semantic theories of truth or reference; for example, against correspondence truth. As a result, much contemporary anti-realist argument is largely beside the realist point. I shall briefly consider two famous examples.[11]

(1) Dummett (1978) identifies realism with an evidence transcendent – in effect, correspondence – view of truth. He goes on to argue that this view is mistaken, that the notion of truth needed in our semantic theory must be an epistemic one based on verification. (2) Putnam has produced a model-theoretic argument (1978, pp. 125–7; 1983, pp. 1–25) against 'metaphysical realism' and in favour of 'internal realism'. Putnam starts by arguing that there cannot be determinate reference relations to a mind-independent reality. As a result, there is no way in which the 'ideal' theory – one meeting all operational and theoretical constraints – could be false. So metaphysical realism is 'incoherent'. The argument has generated a storm of responses.

Now whatever the rights and wrongs of these matters,[12] the arguments have no direct bearing on realism. Dummett's argument is straightforwardly semantic, not metaphysical. Putnam's metaphysical realism is a hybrid of something like realism with something like correspondence truth. The only part of this hybrid that may be directly affected by Putnam's argument about reference is correspondence truth.[13] Indeed, the challenge of Putnam's argument can be posed, and often seems to be posed, in a way that presupposes realism: a representation is related by one causal relation to certain mind-independent entities and by another causal relation to other such entities; which relation determines reference?

I have emphasized that realism is a metaphysical doctrine and hence different from semantic doctrines like correspondence truth. However, realism is a little bit semantic in requiring that the world be independent of our semantic capacities. Similarly, it is a little bit epistemic in requiring that the world be independent of our epistemic capacities. But these are only minor qualifications to the metaphysical nature of realism.

Why has the metaphysical issue been conflated with semantic issues? This is a difficult question but part of the answer is surely the 'linguistic turn' in twentieth-century philosophy. At its most extreme, this turn treats all philosophical issues as being about language (sec. 4.5).

I claim that no semantic doctrine is in any way constitutive of realism (or any metaphysical doctrine of realism). This is not to claim that there is no evidential connection between the two sorts of doctrines. Indeed, I favour the Quinean view that, roughly, everything is evidentially connected to everything else. So distinguishing realism from anything semantic is only the first step in saving it. We have to consider the extent to which contemporary semantic arguments, once conflations are removed, might be used as evidence against realism: although their conclusions do not amount to anti-realism, they may count in favour of anti-realism. Traditionally, philosophers started with an epistemological view and typically used this as evidence against realism. We should reconstrue contemporary philosophers so that they are doing something similar: starting with a semantic view and using it as evidence against realism.

In the next section, I shall assess these arguments against realism, claiming that they start in the wrong place. I shall first consider traditional arguments from epistemology and then, reconstrued contemporary arguments from semantics. The concern here is with realism about observables, common-sense realism. Having established the case for this, I shall argue for scientific realism in the final section.

Why be a common-sense realist?

Realism about the ordinary observable physical world is a compelling doctrine. It is almost universally held outside intellectual circles. From an early age we come to believe that such objects as stones, cats, and trees exist. Furthermore, we believe that these objects exist even when we are not perceiving them, and that they do not depend for their existence on our opinions nor on anything mental. This realism about ordinary objects is confirmed day by day in our experience. It is central to our whole way of viewing the world. Common-sense realism is aptly named because it is the core of common sense.

What, then, has persuaded so many philosophers out of it? A clear answer emerges from the tradition before the linguistic turn (ch. 5). If we have knowledge of the external world, it is obvious that we acquire it through our sensory experiences. Yet, Descartes (1641) asks, how can we rely on these? First, the realist must allow that our senses sometimes deceive us: there are the familiar examples of illusion and hallucination. How, then, can we ever be justified in relying on our senses? Second, how can we be sure that we are not dreaming? Though we think we are perceiving the external world, perhaps we are only dreaming that we are. Finally, perhaps there is a deceitful demon causing us to have sensory experiences 'as if' of an external world, when in fact there is no such world. If we are not certain that this is not the case, how can we know that realism is correct? How can it be rational to believe realism?

One traditional way of responding to the challenge of this extreme Cartesian scepticism is to seek an area of knowledge which is not open to sceptical doubt and which can serve as a 'foundation' for all or most claims to knowledge. Since even the most basic common-sense and scientific knowledge – including that of the existence of the external world – is open to doubt, this search is for a special philosophical realm of knowledge outside science. The foundationalist has always found this realm in the same place. 'In the search for certainty, it is natural to begin with our present experiences' (Russell 1912, p. 1). This natural beginning led traditionally to the view that we could not be mistaken about mental entities called 'ideas'. More recently, it has led to the similar view that we could not be mistaken about entities called 'sense-data'. These entities are 'the given' of experience. I shall talk of 'sense-data'.

From this perspective, the justification of realism can seem hopeless. The perspective yields what is sometimes called, anachronistically, 'the movie-show model' of the mind. Sense-data are the immediate objects of perception. They are like images playing on a screen in the inner theatre of a person's mind. The person (a homunculus really) sits watching this movie and asks herself: (1) Is there anything outside the mind causing the show? (2) If so, does it resemble the images on the screen? To answer these questions 'Yes', as Locke (1690) does with a qualification or two, is to be a 'representative realist'. But Locke's justification for his answers is desperately thin as Berkeley (1710) shows: there seems to be no basis for the inference from the inner show to the external world. Certainly, there is no reason why a Cartesian sceptic should accept the inference.

The problem for realism is the 'gap' between the object known and the knowing mind. According to the realist, the object known is external to the person's mind and independent of it. Yet the person has immediate knowledge only of her own sense-data. She can never leave the inner theatre to compare those sense-data with the external world. So how could she ever know about such a world?

To save our knowledge, it seemed to Berkeley and many others, we must give up realism and adopt idealism: the world is constructed, in some sense, out of sense-data. The gap is closed by bringing objects, one way or another, 'into the mind'. But the problem is that even this desperate metaphysics does not save our knowledge. Idealism too is open to sceptical doubt.

First, consider the foundations of idealism: our allegedly indubitable knowledge of our own sense-data. Why should the sceptic accept that there are any such mental objects as sense-data? Even if there are, why should the sceptic accept that the person has indubitable knowledge of them? Why is this any more plausible than the view that we have indubitable knowledge of external objects?

Even if the foundations are granted, and realism is abandoned, the task of building our familiar knowledge to Cartesian sceptical standards on these foundations has proved impossible.

The simplest part of this knowledge is singular knowledge of physical objects; for example, the knowledge that Nana is a cat. How can we get this knowledge from knowledge of sense-data? This might seem easy if Nana were literally constructed out of sense-data, if she were nothing but a bundle of them. But then how could we explain the fact that Nana can exist unobserved? The obvious answer that sense-data can exist unobserved is quite gratuitous from the sceptical viewpoint.

So idealists favoured a different sort of construction, the 'logical construction' proposed by 'phenomenalism'. Each statement about a physical object was to be translated, in some loose sense, into statements about sense-data. Since the latter statements are the sort that the foundationalist thinks we know, it was hoped in this way to save our knowledge, albeit in a new form. However, the total failure of all attempts to fulfil this translation programme over many years of trying is so impressive as to make it 'overwhelmingly likely' that the programme cannot be fulfilled (Putnam 1975b, p. 20).

From a realist perspective, it is easy to see the problem for phenomenalism: there is a loose link between a physical object and any set of experiences we might have of it. As a result, no finite set of sense-datum statements is either necessary or sufficient for a physical-object statement.

In sum, the foundationalist anti-realist cannot save physical objects. He cannot save even our singular knowledge of the world. We have already noted the failure of foundationalist realism. The Cartesian sceptical challenge leaves the foundationalist no place to stand and no way to move: he is left, very likely, only with the knowledge that he is now experiencing, with 'instantaneous solipsism'. The foundationalist programme is hopeless.[14]

Kant is responsible for another traditional idealist response to the sceptical challenge. Kant's way of saving knowledge is very different from foundationalism's. He closes the gap between the knowing mind and the object known with his view that the object is partly constituted by the cognitive activities of the mind. He distinguishes objects as we know them – stones, trees, cats, and so on – from objects as they are independent of our knowledge. Kant calls the former 'appearances' and the latter 'things-in-themselves'. Appearances are obtained by our imposition of a priori concepts; for example, causality, time, and the Euclidian principles of spatial relations. Only things-in-themselves, forever beyond our ken, have the objectivity and independence required by realism. Appearances do not, as they are partly our construction. And, it must be emphasized, the familiar furniture of the world are appearances not things-in-themselves. Although an idealist, Kant is a 'weak realist' (p. 23).

How does this view help with scepticism? We can know about appearances because, crudely, we make them. Indeed, Kant thinks that we could not know about them unless we made them: it is a condition on the possibility of knowledge that we make them.

Many contemporary anti-realisms combine Kantianism with relativism to yield what is known as 'constructivism'. Kant was no relativist: the concepts imposed to constitute the known world were common to all mankind. Contemporary anti-realisms tend to retain Kant's ideas of things-in-themselves and of imposition, but drop the universality of what is imposed. Instead, different languages, theories, and world-views are imposed to create different known worlds. Goodman, Putnam, and Kuhn are among the constructivists.

Constructivism is so bizarre and mysterious – how could we, literally, make dinosaurs and stars? – that one is tempted to seek a charitable reinterpretation of constructivist talk. But, sadly, charity is out of place here (secs 13.1–3).[15]

Something has gone seriously wrong. The Cartesian sceptical challenge that has persuaded so many to abandon realism has led us to disaster: either to a lack of any worthwhile knowledge or to knowledge at the expense of a truly bizarre metaphysics. It is time to think again.

The disaster has come from epistemological speculations about what we can know and how we can know it. But why should we have any confidence in these speculations? In particular, why should we have such confidence in them that they can undermine realism? Over a few years of living people come to the conclusion that there are stones, trees, cats, and the like, existing largely independent of us. This realism is confirmed day by day in their experience. A Moorean point is appropriate. Realism seems much more firmly based than the epistemological speculations that are thought to undermine it.[16] Perhaps, then, we have started the argument in the wrong place: rather than using the epistemological speculations as evidence against realism, perhaps we should use realism as evidence against the speculations. We should, as I like to say, 'put metaphysics first'.

Indeed what support are these troubling speculations thought to have? Not the empirical support of the claims of science, for that sort of support is itself being doubted. The support is thought to be a priori, as is the support for our knowledge of mathematics and logic. Reflecting from the comfort of armchairs, foundationalists and Kantians decide what knowledge must be like, and from this infer what the world must be like:

a priori epistemology → a priori metaphysics.

The 'Moorean point' alone casts doubt on this procedure and the philosophical method it exemplifies, the a priori method of 'First Philosophy'. But we can do better: the doubt is confirmed by the sorts of considerations adduced by Quine (1952, Introduction; 1953, pp. 42–6). These considerations should lead us to reject a priori knowledge and embrace 'naturalism', the view that there is only one way of knowing, the empirical way that is the basis of science.[17] From the naturalistic perspective, philosophy becomes continuous with science. And the troubling epistemological speculations have no special status: they are simply some among many empirical hypotheses about the world we live in. As such, they do not compare in evidential support with realism. Experience has taught us a great deal about the world of stones, trees, and cats, but rather little about how we know about this world. So epistemology is just the wrong place to start the argument: the sceptical challenge should be rejected. Instead, we should start with an empirically based metaphysics and use that as evidence in an empirical study of what we can know and how we can know it; epistemology itself becomes part of science, 'naturalized epistemology':

empirical metaphysics → empirical epistemology.

And when we approach our metaphysics empirically, realism is irresistable.[18] Indeed, it faces no rival we should take seriously. Thus naturalism supports the Moorean point.

Quine is fond of a vivid image taken from Otto Neurath. He likens our knowledge – our 'web of belief' – to a boat that we continually rebuild whilst staying afloat on it. We can rebuild any part of the boat but in so doing we must take a stand on the rest of the boat for the time being. So we cannot rebuild it all at once. Similarly, we can revise any part of our knowledge but in so doing we must accept the rest for the time being. So we cannot revise it all at once. And just as we should start rebuilding the boat by standing on the firmest parts, so also should we start rebuilding our web.[19] Epistemology is one of the weakest parts to stand on. So also is semantics.

We noted that semantics has been at the centre of contemporary anti-realist arguments. Setting aside the frequent conflation of semantics with metaphysics, I suggested that we reconstrue these arguments as simply offering evidence against realism. So just as traditional philosophers argued for epistemological doctrines that show that we could not know the realist world, we should see Dummett and Putnam as arguing for semantic doctrines that show that we could not refer to the realist world. Since we obviously do know about and refer to the world, the arguments run, the world cannot be realist. The objection to traditional arguments was that they started with a priori speculations on what knowledge must be like and inferred what the world must be like. The objection to contemporary arguments is that they start with a priori speculations on what meaning and reference must be like and infer what the world must be like:

a priori semantics → a priori metaphysics.

From the naturalistic perspective, this uses the wrong methodology and proceeds in the wrong direction. We should proceed:

empirical metaphysics → empirical semantics.

Consider Dummett, for example (ch. 14). His case against realism rests on an argument for a verificationist semantics. This argument rests entirely on claims about linguistic competence, about what meanings we could grasp and what concepts we could have. Why should we believe these claims? They are thought to be known a priori. Naturalism rejects that. As empirical claims their support is very weak, far too weak to threaten something as plausible as realism. Indeed, semantics as a whole is in such a poor state that it is just the wrong place to start in doing metaphysics.

Rather, a realist metaphysics is a firm place to start from – as firm as you could wish for – in doing semantics. With realism as a base, I think the prospects of establishing a non-verificationist semantics based on correspondence truth are promising, although I would be the last to underestimate the dimensions of this task (ch. 6 and secs A.12–15).

Consider Kuhn, for another example (ch. 9). I have noted that Kuhn is a constructivist: he holds that the known world exists only relative to the imposition of concepts by our scientific theories. What drives him to this unlovely metaphysics? Implicitly, a 'meta-induction'[20] against realism along the following lines: past theories posited entitites which, from the perspective of our current theories, we no longer think exist; so, probably, from the perspective of future theories we will come to think that the posits of our present theories do not exist. Kuhn has unobservables primarily in mind but it is important to note that the argument applies even to the familiar observables. Why does Kuhn suppose that, from our current perspective, the posits of past theories do not exist? First, he starts with the semantic issue of whether the terms that purport to refer to those entities really do refer rather than with the metaphysical issue of whether the entities exist. Second, in considering the semantic issue, he takes for granted a 'description' theory of reference. According to this theory, the reference of a term depends on the descriptions (other terms) associated with it in the theory: it refers to whatever those descriptions pick out. Now with theory change, particularly radical theory change, is likely to go the view that those descriptions do not pick anything out. So, from the new perspective, the term in the old theory does not refer. This will be true even of an 'observational' term; think, for example, of descriptions associated with 'the Earth' before the Copernican revolution. So the entities that the old terms purport to refer to do not exist.[21] The objection is, once again, that semantics is the wrong place to start. Set aside until the final section the application of the meta-induction to unobservables. We should be much more confident of the continued existence of familiar observables, despite theory changes, than of any semantic theory. If a description theory of reference counts against that existence, so much the worse for the theory. Many ideas for other theories of reference compatible with realism have emerged in recent times.[22]

The mistaken methodology is reflected in a certain caricature of realism that tends to accompany contemporary anti-realist polemics (sec. 12.6). Thus, according to Putnam, realism requires a 'God's Eye view' (1981, p. 74); that we have 'direct access to a ready made world' (p. 146) and so can compare theories with 'unconceptualized reality' (1979, p. 611); that we can make 'a transcendental match between our representation and the world in itself' (1981, p. 134).[23] According to Richard Rorty the realist believes that we can 'step out of our skins' (1982, p. xix) to judge, without dependence on any concepts, whether theories are true of reality.[24] But, of course, no sane person believes any of this. What realists believe is that we can judge whether theories are true of reality, the nature of which does not depend on any theories or concepts.

What lies behind these views of realism? The answer is clear: the Cartesian picture that leads to the sceptical challenge. According to this picture we are theorizing from scratch, locked in our mental theatres, trying to bridge the gap between our sense-data and the external world. But we are not starting from scratch in epistemology and semantics. We can use well-established theories in physics, biology, and so forth; we already have the entities and relations which those theories posit. And if we were starting from scratch, sceptical doubts would condemn us to instantaneous solipsism. The picture puts the epistemic and semantic carts before the metaphysical horse.

To put the carts back where they belong, we take a naturalistic approach to epistemology and semantics. Reflection on our best science has committed us to the many entities of the largely impersonal and inanimate world. It has not committed us to sense-data and so there is no gap between sense-data and the world to be bridged. We go on to seek empirical explanations of that small part of the world in which there are problems of knowledge and reference: people and language. From the naturalistic perspective, the relations between our minds and reality are not, in principle, any more inaccessible than any other relations. Without jumping out of our skins we can have well-based theories about the relations between, say, Michael and Scottie. Similarly, we can have such theories about our epistemic and semantic relations to Michael and Scottie.

In sum, objections to common-sense realism have come from speculations in epistemology and semantics. The Moorean point is that realism is much more plausible than these speculations; we should put metaphysics first. This point is good on its own but when supported by naturalism it is formidable. From the naturalistic perspective, these speculations cannot be supported a priori and they do not come close to having the empirical support enjoyed by realism. Realism is the only doctrine that can be taken seriously.

Why be a scientific realist?

The argument for scientific realism – realism about the unobservables of science – starts by assuming common-sense realism. And, setting aside some deep and difficult problems in quantum theory, the issue is over the existence dimension, over whether these unobservables exist. For the independence dimension mostly goes without saying once common-sense realism has been accepted.

The basic argument for scientific realism is simple (sec. 7.1). By supposing the unobservables of science exist, we can give good explanations of the behaviour and characteristics of observed entities, behaviour and characteristics which would otherwise remain completely inexplicable. Furthermore, such a supposition leads to predictions about observables which are well confirmed; the supposition is successful.

This argument should not be confused with one version of the popular and much discussed argument that 'realism explains success' (sec. 7.3).[25] This version is most naturally expressed talking of truth. First we define success: for a theory to be successful is for its observational predictions to be true. Why is a theory thus successful? The realist argument claims: because the theory is true. However, given the conflation of realism with correspondence truth criticized in the first section, it is worth noting that this talk exploits only the disquotational property of 'true' and so does not require any robust notion of truth. This can be seen by rewriting the explanation without any talk of truth at all. Suppose a theory says that S. The rewrite defines success: for this theory to be successful is for the world to be observationally as if S. Why is the theory thus successful? The rewrite claims: because S. For example, why is the world observationally as if there are atoms? Why are all the observations we make just the sort we would make if there were atoms? Answer: because there are atoms. This realist explanation has a trivial air to it because it is only if we suppose that there are not x's that we feel any need to explain why it is as if there are x's. Still, it is a good explanation. And the strength of scientific realism is that the anti-realist has no explanation of this success: if scientific realism were not correct, realists are fond of saying, it would be 'a miracle' that the observable world is as if it is correct.

This popular argument is good but it is different from my simple one and not as basic. Where the popular argument uses realism to explain the observational success of theories, my simple one uses realism to explain the observed phenomena, the behaviour and characteristics of observed entities. This is not to say that observational success is unimportant to the simple argument: the explanation of observed phenomena, like any explanation, is tested by its observational success. So according to the simple argument, scientific realism is successful; according to the popular one, it explains success. There is not even an air of triviality about the simple argument.

I shall conclude by briefly considering three arguments against scientific realism. (1) The first is an influential empiricist argument. Richard Boyd, who does not agree with its conclusion, has nicely expressed the argument as follows:

> Suppose that *T* is a proposed theory of unobservable phenomena ... A theory is said to be empirically equivalent to *T* just in case it makes the same predictions about observable phenomena that *T* does. Now, it is always possible, given *T*, to construct arbitrarily many alternative theories that are empirically equivalent to *T* but which offer contradictory accounts of the nature of unobservable phenomena ... *T* and each of the theories empirically equivalent to it will be equally well confirmed or disconfirmed by any possible observational evidence ... scientific evidence can never decide the question between theories of unobservable phenomena and, therefore, knowledge of unobservable phenomena is impossible. (1984, pp. 42–4)

One way of putting this is: we should not believe *T* because it is underdetermined by the possible evidence. Commitment to the existence of the entities posited by *T*, rather than merely to the pragmatic advantages of the theory that talks of them, makes no evidential difference, and so is surely a piece of misguided metaphysics; it reflects super-empirical values, not hard facts.

Talk of 'possible evidence' is vague (sec. 3.5). If it is construed in a restricted way then theories may indeed be underdetermined by the possible evidence. Yet for underdetermination to threaten scientific realism, I argue (sec. 7.4), the talk of 'possible evidence' must be construed in a very liberal way. And construed in this way, there is no reason to believe in underdetermination.

One sense of 'possible evidence' (see Quine 1970, p. 179; van Fraassen 1980, pp. 12, 60, 64) is restricted in that it does not cover anything non-actual except acts of observation: it is restricted to all the points of actual space-time that we would have observed had we been around. Yet there are many things that we do not do, but could do, other than merely observing. If we had the time, talent, and money perhaps we could invent the right instruments and conduct the right experiments to discriminate between *T* and its rival *T'*. There may be many differences between the theories which we would not have detected if we had passively observed each point of actual space-time but which we would have detected if we had actively intervened (Hacking 1983) to change what happened at points of space-time. In this liberal sense that allows for our capacity to create phenomena, the class of possible evidence seems totally open.

In the light of this, given any T, what possible reason could there be for thinking a priori that we could not distinguish it empirically from any rival if we were ingenious enough in constructing experiments and auxiliary hypotheses? It is of course possible that we should be unable to distinguish two theories: we humans have finite capacities. The point is that we have no good reason for believing it in a particular case. Even less do we have a good reason for believing it in all cases; that is to say, for believing that every theory faces rivals that are not detectably different.

Behind these realist doubts about underdetermination lies the following picture. T and T' describe different causal structures alleged to underlie the phenomena. We can manipulate the actual underlying structure to get observable effects. We have no reason to believe that we could not organize these manipulations so that, if the structure were as T says, the effects would be of one sort, whereas if the structure were as T' says, the effects would be of a different sort.

If the liberally interpreted underdetermination thesis were true, realism might be in trouble. But why should the realist be bothered by the restricted thesis? A consequence of that thesis is that we do not, as a matter of fact, ever conduct a crucial experiment for deciding between T and T'. This does not show that we could not conduct one. And the latter is what needs to be established for the empiricist argument against realism (Boyd 1984, p. 50). The restricted empirical equivalence of T and T' does not show, in any epistemologically interesting sense, that they make 'the same predictions about observable phenomena', nor that they 'will be equally well confirmed or disconfirmed by any possible evidence'. It does not show that 'scientific evidence can never decide the question between theories of unobservable phenomena and [that], therefore, knowledge of unobservable phenomena is impossible'. It does not show that commitment to T rather than T' is super-empirical and hence a piece of misguided metaphysics.

(2) van Fraassen (1980, 1985) has proposed a doctrine he calls 'constructive empiricism'. It is common-sense but not scientific realist. Suppose that a theory says that S. van Fraassen holds that we may be justified in believing that the observable world is as if S but we are never justified in believing that S. So scientific realism is unjustified. From the realist perspective, such a position amounts to an unprincipled selective scepticism against unobservables: it offends against unobservable rights.[26] An epistemology that justifies a belief in observables will also justify a belief in unobservables. An argument that undermines scientific realism, will also undermine common-sense realism.

So the realist has a simple strategy against such anti-realism. First, she demands from the anti-realist a justification of the knowledge that she claims to have about observables. Using this she attempts to show, positively, that the epistemology involved in this justification will also justify knowledge of unobservables. Second, she attempts to show, negatively, that the case for scepticism about unobservables produced by the anti-realist is no better than the case for scepticism about observables. I claim that arguments along these lines work against van Fraassen (ch. 8).[27]

(3) Finally, perhaps the most influential recent argument against scientific realism arose from the revolution in the philosophy of science led by Kuhn (1962). It is the earlier-mentioned 'meta-induction': past theories posited entities which, from the perspective of our current theories, we no longer think exist; so, probably, from the perspective of future theories we will come to think that the posits of our present theories do not exist. In the second section I argued that the case offered for the premise of this meta-induction rests on two mistakes: first, the mistake of putting semantics before metaphysics; second, the mistake of taking a description theory of reference for granted. This is enough to remove concern about the existence of past observables, but not of past unobservables. For, even without these mistakes there is plausibility to the idea that we no longer believe in the existence of past unobservables; phlogiston is a popular example. The meta-induction against scientific realism is a powerful argument. Still, I think that the realist has a number of defences against it which are jointly sufficient (sec. 9.4).

In conclusion, I have argued that the metaphysical issue of realism about the external world is quite distinct from semantic issues about truth. Furthermore, we should not follow the tradition and argue the metaphysical issue from a perspective in epistemology, nor follow the recent linguistic turn and argue it from a perspective in semantics. Rather, we should adopt naturalism and argue the metaphysical issue first. When we do, the case for common-sense realism is overwhelming and the case for scientific realism is very strong.

The realism dispute arises from the age-old metaphysical question, 'What ultimately is there, and what is it like?' I am sympathetic to the complaint that realism, as part of an answer to this question, is rather boring. Certainly it brings no mystical glow. Nevertheless, it needs to be kept firmly at the front of the mind to avoid mistakes in theorizing about other, more interesting, matters in semantics and epistemology where it makes a difference.[28]

Notes

1 Susan Haack (1987) distinguishes nine 'senses' of 'realism'.
2 Some other examples: Hesse (1967, p. 407); Hooker (1974, p. 409); Papineau (1979, p. 126); Ellis (1979, p. 28); Boyd (1984, pp. 41–2); Miller (1987); Fales (1988, pp. 253–4); Jennings (1989, p. 240); Matheson (1989); Kitcher (1993); Brown (1994).
3 Two examples are Putnam's 'metaphysical realism' (1978, pp. 123–5), and the account of realism by Arthur Fine (1986a, pp. 115–16, 136–7).
4 For fairly accessible accounts of these worlds see, respectively: Kant (1783); Dummett (1978, preface and chs 10 and 14); Goodman (1978); Putnam (1981); Kuhn (1962). In characterizing the independence of the paradigm realist objects, stones, trees, cats, and the like, we deny that they have any dependence on us except the occasional familiar causal one. Other physical objects that have a more interesting dependence on us – for example, hammers and money – pose more of a challenge to the characterization. But, with careful attention to the differences between this sort of dependence and the dependence that anti-realists allege, the challenge can be met (secs 13.5–7).
5 Cf. Putnam (1985, p. 78; 1987, pp. 15–16). Most philosophers who tie realism to correspondence truth do not argue for their position. Dummett is one exception, criticized in my ch. 14. Michael Williams (1993, p. 212n) is another, criticized in my sec. A.2.
6 Note that the point is not a verbal one about how the word 'realism' should be used. The point is to distinguish two doctrines, whatever they are called (p. 40).
7 More needs to be said to allow for the paradoxes, ambiguity, indexicals, and truth value gaps.
8 The utility of 'true' that comes from its disquotational property is much greater than the examples in this paragraph show. On this, and the idea of deflationary truth, see my sec. 3.4 and the works it draws on.
9 Some will object that we cannot assess realism until we have interpreted it and this requires a semantic theory that talks of truth. I argue against this objection in secs 4.6–9, A.2–11.
10 See, for example, Leeds (1978), Field (1978), Churchland (1979), Stich (1983).
11 Two other examples are: Rorty (1979), discussed in my ch. 11; Laudan (1981), discussed in my ch. 9.
12 I argue that Dummett is wrong in ch. 14, and that Putnam is in ch. 12 and secs A.16–19.
13 Putnam criticizes other views that he associates with metaphysical realism and that are also inessential to realism. One example is the view that there is exactly one true and complete description of the world (1981, p. 49), a view which, with correspondence truth, is alleged to require 'a ready made world' (1983, p. 211; cf my sec. 13.4). Another example is a sort of individualistic essentialism (1983, pp. 205–28). Even if Putnam's criticisms of these views were correct, they would leave realism largely untouched.
14 Note that the programme we are talking about attempts to answer the Cartesian sceptic by rebuilding our knowledge on the foundation of indubitable knowledge of sense-data, mental entities that are the immediate objects of perception. Less demanding forms of foundationalism that do not make this attempt to answer the sceptic may well be promising; see note 18.

15 Because constructivism is so bizarre and mysterious, its popularity cries out for explanation. I have tried to offer some rational explanations (secs 13.4–7). For some learned, and very entertaining, explanations of a different sort, see Stove (1991). Stove thinks that anti-realism, like religion, stems from our need to have a congenial world. For some suggestions by Georges Rey along similar lines, see my p. 257, n. 11.

16 Steven Hales drew my attention to the Moorean nature of this point. Note that the point is not that realism is indubitable, to be held 'come what may' in experience: that would be contrary to naturalism. The point is that, prima facie, there is a much stronger case for realism than for the speculations. (Thanks to Paul Boghossian.)

17 A particularly important consideration against the a priori, in my view (1996, sec. 2.2), is that we lack anything close to a satisfactory explanation of a non-empirical way of knowing. We are told what this way of knowing is not – it is not the empirical way of deriving knowledge from experience – but we are not told what it is. Rey (1998) and Field (1998) have a more tolerant view of the a priori. My (1998) is a response.

18 Some people think that science itself undermines realism. I think that this is a mistake (secs 5.10, 7.9).

19 It is plausible to think that the firmest parts are our singular beliefs about the objects we observe. So we might hope for a new foundationalism built on these beliefs, one with no pretensions to answer the unanswerable Cartesian sceptic, and with no presumption that the beliefs are indubitable.

20 This apt term, and formulations of the argument along these lines, are due to Putnam (1978, p. 25).

21 A similar line of argument has been used by Stich (1983) and others to argue for various forms of eliminativism about the mind. Happily, Stich has recently recanted (1996, pp. 3–90).

22 See, for example, Kripke (1980), Putnam (1975), Dretske (1981), and Millikan (1984).

23 Putnam attributes this view to realist friends 'in places like Princeton and Australia' (1979, 611). The Dummettians have more bad news for Australians (particularly black ones): 'there is no sense to supposing that [Australia] either determinately did or did not exist [in 1682]' (Luntley 1988, pp. 249–50).

24 See also (1979, p. 293); Fine (1986a, pp. 131–2; 1986b, pp. 151–2).

25 I identify eight versions of the argument by distinguishing different senses of 'realism' and 'success' (sec. 6.6).

26 This, not the legitimacy of 'abduction', is the primary issue in the defence of scientific realism; cf. Laudan (1981, p. 45); Fine (1986a, pp. 114–15; 1986b, p. 162).

27 However, it should be noted that my discussion does not take account of van Fraassen's radical non-justificationist epistemology (1989).

28 My thanks to Steven Hales and Georges Rey for comments on a draft of this chapter. Slightly revised version; the chapter is reprinted from *Metaphysics: Contemporary Readings* (1998), with permission of Wadsworth.

32   *Michael Devitt*

References

Berkeley, George (1710), *Principles of Human Knowledge*.

Boyd, Richard N. (1984), 'The Current Status of Scientific Realism', in Leplin 1984a, pp. 41–82.

Brown, James Robert (1994), *Smoke and Mirrors: How Science Reflects Reality*, New York, Routledge.

Churchland, Paul M. (1979), *Scientific Realism and the Plasticity of Mind*, Cambridge, Cambridge University Press.

Descartes, René (1641), *Meditations on First Philosophy*.

Devitt, Michael (1996), *Coming to Our Senses: A Naturalistic Defense of Semantic Localism*, New York, Cambridge University Press.

Devitt, Michael (1997), *Realism and Truth*, Princeton, Princeton University Press, 2nd ed. with a new Afterword (1st ed. 1984, 2nd ed. 1991).

Devitt, Michael (1998), 'Naturalism and the A Priori', *Philosophical Studies*, 92, pp. 45–65.

Dretske, Fred I. (1981), *Knowledge and the Flow of Information*, Cambridge, MA, MIT Press.

Dummett, Michael (1978), *Truth and Other Enigmas*, Cambridge, MA, Harvard University Press.

Ellis, Brian (1979), *Rational Belief Systems*, Oxford, Basil Blackwell.

Fales, Evan (1988), 'How to be a Metaphysical Realist', in *Midwest Studies in Philosophy, Volume XII: Realism and Anti-realism*, Peter A. French, Theordore E. Uehling, Jr., and Howard K. Wettstein (eds), Minneapolis, University of Minnesota Press, pp. 253–74.

Field, Hartry (1978), 'Mental Representation', *Erkenntnis*, 13, pp. 9–61.

Field, Hartry (1998), 'Epistemological Nonfactualism and the A Prioricity of Logic', *Philosophical Studies*, 92, pp. 1–24.

Fine, Arthur (1986a), *The Shaky Game: Einstein, Realism, and the Quantum Theory*, Chicago, University of Chicago Press.

Fine, Arthur (1986b), 'Unnatural Attitudes: Realist and Instrumentalist Attachments to Science', *Mind*, 95, pp. 149–77.

Goodman, Nelson (1978), *Ways of Worldmaking*, Indianapolis, Hackett Publishing Company.

Haack, Susan (1987), 'Realism', *Synthese*, 73, pp. 275–99.

Hacking, Ian (1983), *Representing and Intervening: Introductory Topics in the Philosophy of Natural Science*, Cambridge, Cambridge University Press.

Hesse, Mary (1967), 'Laws and Theories', in *The Encylopedia of Philosophy*, Paul Edwards (ed.), New York, Macmillan, vol. 4, pp. 404–10.

Hooker, Clifford A. (1974), 'Systematic Realism', *Synthese*, 51, pp. 409–97.

Jennings, Richard (1989), 'Scientific Quasi-Realism', *Mind*, 98, pp. 223–45.

Kant, Immanuel (1783), *Prolegomena to Any Future Metaphysics*.

Kitcher, Philip (1993), *The Advancement of Science: Science Without Legend, Objectivity Without Illusions*, New York, Oxford University Press.

Kripke, Saul A. (1980), *Naming and Necessity*, Cambridge, MA, Harvard University Press.

Kuhn, Thomas S. (1962), *The Structure of Scientific Revolutions*, Chicago, Chicago University Press, 2nd ed. 1970.

Laudan, Larry (1981), 'A Confutation of Convergent Realism', *Philosophy of Science*, 48, pp. 19–49. Reprinted in Leplin 1984a.

Leeds, Stephen (1978), 'Theories of Reference and Truth', *Erkenntnis*, 13, pp. 111–29.

Leplin, Jarrett (ed.) (1984a), *Scientific Realism*, Berkeley, University of California Press.

Leplin, Jarrett (1984b), 'Introduction', in Leplin 1984a, pp. 1–7.

Locke, John (1690), *An Essay Concerning Human Understanding.*

Luntley, Michael (1988), *Language, Logic and Experience: The Case for Anti-Realism*, La Salle, Open Court.

Matheson, Carl (1989), 'Is the Naturalist Really Naturally a Realist?', *Mind*, 98, pp. 247–58.

Miller, Richard W. (1987), *Fact and Method: Explanation, Confirmation and Reality in the Natural and Social Sciences*, Princeton, Princeton University Press.

Millikan, Ruth (1984), *Language, Thought, and Other Biological Categories: New Foundations for Realism*, Cambridge, MA, MIT Press.

Papineau, David (1979), *Theory and Meaning*, Oxford, Clarendon Press.

Putnam, Hilary (1975), *Mind, Language and Reality, Philosophical Papers*, vol. 2, Cambridge, Cambridge University Press.

Putnam, Hilary (1978), *Meaning and the Moral Sciences*, London, Routledge and Kegan Paul.

Putnam, Hilary (1979), 'Reflections on Goodman's Ways of World-Making', *Journal of Philosophy*, 76, pp. 603–18.

Putnam, Hilary (1981), *Reason, Truth and History*, Cambridge, Cambridge University Press.

Putnam, Hilary (1983), *Realism and Reason: Philosophical Papers*, vol. 3, Cambridge, Cambridge University Press.

Putnam, Hilary (1985), 'A Comparison of Something with Something Else', *New Literary History*, 17, pp. 61–79.

Putnam, Hilary (1987), *The Many Faces of Realism*, LaSalle, Open Court.

Quine, W. V. O. (1952), *Methods of Logic*, London, Routledge and Kegan Paul.

Quine, W. V. O. (1953), *From a Logical Point of View*, Cambridge, MA, Harvard University Press, 2nd ed., rev., 1st ed., 1953.

Quine, W. V. (1970), *Philosophy of Logic*, Cambridge, MA, Harvard University Press.

Rey, Georges (1998), 'A Naturalistic A Priori', *Philosophical Studies*, 92, pp. 25–43.

Rorty, Richard (1979), *Philosophy and the Mirror of Nature*, Princeton, Princeton University Press.

Rorty, Richard (1982), *Consequences of Pragmatism (Essays: 1972–1980)*, Minneapolis, University of Minnesota Press.

Russell, Bertrand (1912), *The Problems of Philosophy*, London, Oxford Paperbacks, 1967. Original publ. 1912.

Stich, Stephen P. (1983), *From Folk Psychology to Cognitive Science: The Case Against Belief*, Cambridge, MA, MIT Press.

Stove, David (1991), *The Plato Cult and Other Philosophical Follies*, Oxford, Basil Blackwell.

van Fraassen, Bas C. (1980), *The Scientific Image*, Oxford, Clarendon Press.

van Fraassen, Bas C. (1985), 'Empiricism in the Philosophy of Science', in *Images of*

34 *Michael Devitt*

Science: Essays on Realism and Empiricism, with a Reply from Bas C. van Fraassen, Paul M. Churchland and Clifford A. Hooker (eds), Chicago, University of Chicago Press, pp. 245–308.

van Fraassen, Bas C. (1989), *Laws and Symmetry*, Oxford, Clarendon Press.

Williams, Michael (1993), 'Realism and Scepticism', in *Reality, Representation, and Projection*, John Haldane and Crispin Wright (eds), New York, Oxford University Press, pp. 193–214.

Chapter Three

Metaphysical and Scientific Realism

Evandro Agazzi

Discussion of realism and anti-realism is very widespread in contemporary philosophy. But this debate has not proven very fruitful, for the simple reason that little agreement can be found on the meaning of the fundamental concepts involved. These differences are not just a matter of personal idiosyncrasies, but are rather the often unconscious consequence of a mixture of different-meaning elements that have been attached to realism during the whole history of Western philosophy. For this reason, it will be useful briefly to summarize the salient features of this historical development.

Realism in 'classical' philosophy

The notion of realism has had two basic meanings in the history of Western philosophy: the first emerged in the 'dispute about universals' of the Middle Ages and concerned the 'kind of existence' that can be attributed to universals, such as genera and species. It is, therefore, an ontological question. In this dispute no one denied that 'full reality' should be attributed to so-called 'individual substances' that 'exist in themselves', such as stones, trees, men, and women, but also God, angels, and devils. The issue concerned only the real existence of such 'abstract entities' as genera and species, that we commonly denote through general concepts (this is why the dispute was called the 'dispute about universals'). Given this premiss, it has become customary to call 'extreme realists' those philosophers who claimed that universals are 'real', that is, have an existence in themselves, as Plato had said of his Ideas. A different position was advocated by the so-called 'conceptualists' who affirmed that universals are simply concepts and, as such, though endowed with some 'kind of reality', they do not exist 'in themselves', but only in our minds. They have simply a mental reality, that is, a mental existence: they are *entia rationis*, in the terminology of that time. A radical position was taken by the 'nominalists', who were prepared to grant real existence only to individuals. Nominalists reduced the 'universals' to simple 'names' to which not even a conceptual designatum corresponds: they are only mental (and linguistic) tools for grouping together individuals that show certain similarities. Finally, a fourth position was that of the 'moderate realists' inspired by Aristotelian metaphysics (the most famous representative of which is Thomas Aquinas). According to them universals do not exist 'in themselves', but exist 'in the individual substances' as their 'essence' or 'form' (in the technical Aristotelian sense), as well as 'in our intellect' as concepts.

These different positions were all of a strict ontological or metaphysical kind, and were not rooted in different epistemological doctrines. Indeed, all tacitly shared a common realist epistemology in the following sense: our knowledge is knowledge of reality in its different kinds. Moreover, all admitted that our knowledge is based on sensible intuition and on intellectual intuition, and differed only in the determination of the objects of such intuitions. The extreme realists maintained that genera and species have an ontological subsistence in themselves as immaterial substances, and can be grasped directly by our intellectual intuition. Moderate realists maintained that these universals have no subsistence in themselves, but enter into the ontological constitution of the individual substances as their form, and can be grasped by our intellectual intuition by means of 'abstraction'. Therefore, they have a kind of double existence, or ontological status: on the one hand, they are the ontological constituents that, in conjunction with matter, make up the individual substances; on the other hand, they are present in our intellect as *entia rationis*. The conceptualists limited their position to this second aspect, without feeling committed to a particular doctrine regarding the status of the universals 'outside' the mind. The nominalists considered the intellectual intuition as a kind of reproduction of the sensible intuition, which is, admittedly, intuition only of single individual items of reality and, as a consequence, could not give the universals any consistent ontological status. In conclusion, the different positions can be related to different ways of considering the role of intellectual intuition.

We cannot delay here over the details of the 'analytical ontology' that provided the ground for such discussions. We simply want to stress that a 'realist' attitude was their common framework. This may become even more apparent if we consider that in such ontological disputes the famous issue of 'independence' was also touched upon, but this was by no means conceived as an independence of existence from knowledge. For example, the existence of accidents was said to 'depend' on the existence of the substance in which they inhere, but not on the activity of the mind that perceives them.

Epistemological dualism

A radical change emerged from a tacit and gratuitous presupposition that characterized 'modern' philosophy (conventionally inaugurated by Descartes), according to which what we immediately know are our representations or ideas, and not 'reality'. Therefore the chief question became that of knowing whether or not, starting from our ideas, we can indirectly attain knowledge of reality. Those who maintained that we can were known as 'realists', while those who maintained that we are condemned to know simply our ideas were known as 'idealists'. Therefore, 'realism' has, in this context, an epistemological meaning. It is obvious that this tacit 'presupposition' actually contained a second: that 'reality' exists independently of our knowledge (a presupposition that can be called 'naturalistic' since it reflects the common-sense conviction that there is a Nature 'external' to our mind, and at the same time constitutes the presupposition of natural science, whose spontaneous aim is that of investigating and knowing the features of this Nature). As a consequence the majority of modern philosophers until the end of the eighteenth century can be qualified as 'realist' in the weak ontological sense of admitting (in keeping with common sense) that there exists a reality 'external' to our mind, whose existence is independent of our mind (external things are not created or posited by our mind); while they were also 'idealist' in the weak epistemological sense of affirming that we do not directly know reality, but only our ideas. Therefore the open question remained that of establishing whether or not we can also attain an (indirect) knowledge of reality starting from our ideas, that is, whether or not we can know how reality is, what it is like. From this second point of view, the realists are those philosophers who believed that a positive answer may be given to this question, and idealists are those who did not see the possibility of transcending the realm of our ideas to reach reality. Before the end of the eighteenth century only Berkeley, among the best-known philosophers, expressed the full and radical idealist position, in which the 'ontological dependence' of things on our ideas is explicitly affirmed: *esse est percipi* ('to be is to be the content of a perception'). In the nineteenth century the so-called 'transcendental idealism' of Fichte, Schelling and, especially, Hegel, arrived at the extreme conclusion of claiming the 'ontological identity' of reality and thought, in the sense that reality 'reduces' to thought.

All these are quite well-known facts, and we have recalled them only to stress that they are a consequence of the 'dualistic presupposition' mentioned before. The interesting point is to state the crucial difference that divides 'classical' from 'modern' epistemology. This difference can be summarized as follows: according to classical epistemology knowledge consists in the fact that things are present to the mind; according to modern epistemology knowledge consists in the fact that things are only (at best) represented by the mind. Moreover, this presence was not to be conceived of in any spatial sense, and this conception was expressed through the notion of a particular 'identity', the intentional identity of thought and reality: in a perception or in an intellectual intuition our cognitive capacities 'identify' themselves with the objects, though remaining ontologically distinct from them. This ontological distinction furnishes the correct meaning of the 'external' world which, otherwise, would reduce to the almost ridiculous notion of all that lies 'outside of my skin'. The representation, from this point of view, simply is 'the way of being present' of a given thing to our cognitive capacities, and 'depends' in an ontological sense on both, but not in the sense of being 'produced' by either of them.

Modern epistemology, having lost the notion of 'intentional identity', gives representations the status of direct objects of knowledge that we encounter in our mind. Therefore a spontaneous question is: from where do our ideas come? This famous problem of the 'origin of ideas' has concerned many modern philosophers, but is as gratuitous as the dualistic presupposition itself. Indeed, why should we believe that our ideas 'come from' anything else? The answer to this ill-posed question was sought along two equally unsatisfactory paths of causal explanation. The so-called 'rationalists' maintained that our basic ideas are innate and are put into our mind (they are caused) directly by God. The 'empiricists' maintained that our basic ideas are causally produced by a physical action of things on our sense organs. Both proposals had their peculiar difficulties, that are presented in any textbook of the history of philosophy, and need not be recalled here. It is worth noting that the conviction that our representations could with luck be a copy of, or have a 'similitude' with other things is a typical consequence of the 'epistemological presupposition' and characterizes, therefore, modern philosophy, while it cannot be correctly attributed to classical epistemology, according to which things, and not their alleged copies, were intentionally present in the mind.

Metaphysical realism

The expression 'metaphysical realism' is known for having been used by Putnam, who means by it the philosophical perspective according to which

> the world consists of some fixed totality of mind-independent objects. There is exactly one true and complete description of 'the way the world is'. Truth involves some sort of correspondence relation between words or thought-signs and external things and sets of things.

He calls this perspective 'the *externalist* perspective, because its favourite point of view is a God's Eye point of view'.[1] It may seem a little mysterious that such a perspective should be called realist in a 'metaphysical' sense. But we can venture to propose an historical explanation of this fact. The term 'metaphysics' in the 'classical' sense had two distinct shades of meaning (which are already present in Aristotle): on the one hand it designated the doctrine of 'reality as such' (that is, of the most general characteristics of being); on the other hand it designated the doctrine of those dimensions of reality that transcend the domain of sensible experience. Logical empiricists of the Vienna Circle were programmatically against metaphysics, which they intended concretely in the second sense (for them, any discourse transcending sense experience is 'meaningless'). Analytical philosophy, however, which to a certain extent can be seen as derived from logical empiricism, gradually recovered at least in part the first meaning of classical metaphysics, and legitimated a so-called 'descriptive metaphysics', which is intended to be a study of the general features of empirically accessible reality (leaving aside those developments that had led classical metaphysics to affirm the existence, and describe some characteristics, of meta-empirical or transcendent reality). It is not without interest that some of the thinkers who have worked in the framework of this descriptive metaphysics have gradually rediscovered several concepts and principles of Aristotelian metaphysics (Strawson and Wiggins are good examples). This position can be characterized by two 'realist' attitudes: 'ontological realism' (reality exists in itself and is independent of our knowledge of it), and 'epistemological realism' (we are able to know what reality is like or 'the way the world is'). If we focus our attention on these two basic features, Putnam's concept of 'metaphysical realism' does not appear mysterious, for it can be related to a conception and terminology that are significantly present in contemporary English language philosophy. Yet it is doubtful that the additional claims he attributes to the metaphysical realist are correctly attributed to him. In fact he adds to the usual claim that the world consists of some totality of mind-independent objects (ontological requirement), the gratuitous pretension that this totality is 'fixed' (without explaining what this actually means). The claim that 'there is just one true and complete description of the way the world is' is again excessive, because it presupposes that the metaphysical realist necessarily advocates a one-to-one correspondence between elements, properties, relations of the world and fixed terms of the language (which is, at best, a rough form of logical atomism). Finally the epistemological claim is made that the metaphysical realist would identify adequate knowledge of reality with the 'God's Eye point of view'. If all these features were joined together it would be hard to find a single 'metaphysical realist' in the history of philosophy. It would not be difficult to show that these additional alleged requirements depend on a conflation of ontological, epistemological, semantic, and linguistic issues that, though often being mixed up in discussions of realism, ought to be carefully distinguished. We are not interested, however, in criticizing Putnam now, since we are more interested in proposing what we believe to be a more adequate definition of metaphysical realism.

We propose to call 'metaphysical realism' simply the position that does not subscribe to the two uncritical and dogmatic presuppositions mentioned above, that is, that we directly know only our representations, and that there is a reality hidden behind these representations. No evidence or argument has ever been proposed for such claims, and we can, on the contrary, maintain that reality is present to our senses and thinking. The burden of proof lies with those who claim that what is present is not reality, but something else. It is not easy to see how they could substantiate it. Indeed, unless one conceives of reality in a naive and purely pictorial sense, as something mysterious lying behind what is present to our knowing capability, reality simply means the totality of what is real, and real is, in turn, anything that is different from nothing. Therefore, to say that thinking or knowing is not thinking or knowing reality amounts to saying that we think or know nothing, and this amounts to having no thinking or knowledge at all. All this does not entail that the whole of reality is present in every act of knowledge. On the contrary, only certain aspects or attributes of reality are present in our acts of knowledge (only colours and shapes can be present to our visual perceptions, and not sounds or smells, for example; moreover, only a single colour and a single shape are present in a single accurately focused act of vision). Since human beings are endowed with cognitive capabilities that overstep those of other living beings (let us call them intellectual capabilities for brevity), it is absolutely obvious that certain features or attributes of reality can be present to our intellect, and these are those features that we call universal or abstract. If we call 'intuition' the immediate presence of something to our cognitive capabilities, we must conclude that, besides sensible intuition, we are endowed also with intellectual intuition.

The characterization of realism we have defended thus far can be summarized as the thesis of the 'knowability and intelligibility of reality'. This does not mean, we repeat, that the whole of reality is actually known or even totally accessible to our knowledge. Therefore this realism is external in Putnam's sense, but does not coincide at all with a 'God's Eye view'. Why then do we call it 'metaphysical'? The reason is that, while admitting that what is immediately known is real, is a part of reality, we admit that the domain of reality is broader than the domain of what is immediately present to our knowledge, since at the same time we recognize that the intellect can lead us to determine features of reality that are not immediately present. Here, again, we are not maintaining that 'the whole of reality' can be disclosed to us, but simply that more of reality can be known than what is immediately present to our sensible and intellectual intuition. This can be obtained by logical inference that relies upon the basic intelligibility of reality and justifies the admission of features of it that are necessary for making it intelligible or, at least, that are sufficient for this purpose (with familiar differences in degrees of certainty). If we call 'experience' the whole of what is immediately present (that is the content of sensible and intellectual intuition), this enterprise corresponds to a transcendence of experience and, since this move is the core of any metaphysics conceived without aprioristic limitations, this explains why the denomination of 'metaphysical realism' is appropriate.

Our claim is that metaphysical realism is not a philosophical position limited to the rational construction of metaphysics in a disciplinary sense, but is a kind of methodological prerequisite for the construction of science as well and, in particular, justifies a scientific realism that is at the same time 'internal' and 'external' in Putnam's sense. It is 'internal' in the sense that only those features of reality are investigated that fall 'within' the criteria of reference, the conceptual schemes, the methodological procedures admitted by a given science, but is also 'external' in the sense that all these 'cognitive tools' do not 'pose' or 'construct' in an ontological sense those attributes of reality to which they refer. Such referents are 'external' to scientific knowledge in an ontological sense.

Evandro Agazzi

Historical overview

Modern science was 'realist' from its origins. A non-realist view of it was advocated by Kant for very special reasons, but did not really affect the common appreciation of science until the 'crisis' of the exact sciences that occurred at the end of the nineteenth century. We will briefly substantiate these claims.

The principal reason why Galileo can be rightly considered as the founder of modern science is that he explicitly and consciously stated the conditions for attaining effective knowledge of 'natural substances' (that is, of physical bodies). Instead of trying to 'grasp by speculation the intimate essence' of such substances (which was the condition traditionally required for scientific knowledge in general, but which he declared to be a 'desperate enterprise'), he maintained that we can attain knowledge of 'some of their affections'. This programme is still expressed in the language of classical ontology, in which a 'substance' was characterized by an 'essence' and the possession of certain 'accidents', of which 'affections' represented a particular kind. All this belongs to a realist ontology. Among the accidents of physical bodies Galileo distinguished those that depend on the sensory capabilities of the observer (colours, smells, and so on), and are therefore subjective, from those that are intrinsic to the body (they are the quantifiable and mathematizable qualities), and that he calls, for this reason, 'real accidents'. It is only with these real accidents that natural science is concerned, and it can do this efficaciously by adopting mathematics as a means for describing them. This clearly realist view of natural science is abundantly confirmed throughout Galileo's works, and in particular by his refusal to consider Copernican astronomy simply as a means for more easily 'saving the appearances' of celestial phenomena, rather than as a successful effort to determine the 'real constitution of the universe'.[2] What can be correctly found in the Galilean assertions is a certain weakening of the force of intellectual intuition (it is no longer credited with the capacity to capture the essence of things). Its role, however, remains primary, since it is only thanks to intellectual intuition that mathematical properties can be determined and described, that mathematical models of physical events can be constructed, and that idealizations of the natural phenomena can be proposed, and these are salient characteristics of Galileo's scientific method. Therefore, he cannot be considered an empiricist because, while declaring that natural science is based on 'sensible experiences and mathematical demonstrations', he also admits that the most significant advances occur when 'the intellect does violence to the senses'.

In the works of Galileo the term 'phenomenon' does not occur, while it is frequently used by Newton. It must be clearly said, however, that Newton's concept of phenomenon is not affected by the 'epistemological dualism' we have described. For him phenomena are simply the 'manifest' characteristics of physical events, and are by no means 'pure appearances'. He simply takes as a basic methodological requirement for what he calls 'natural philosophy' or 'experimental philosophy' the abstention from introducing in this philosophy (that is, in natural science) any 'hidden qualities' that (as he says) traditional philosophers used to posit as contained in the 'substantial forms' of things, in order to provide explanations of the manifest features of things. All this is well in keeping with the views of Galileo (whom Newton mentions with approval on several occasions). With Galileo he also shares the admission of a limited role for intellectual intuition, to the extent that he too recognizes the decisive importance of mathematization in the construction of natural science, but he is much more clearly an empiricist, since the single general laws of physical phenomena are explicitly declared by him to be propositions obtained by 'inductive generalizations' over phenomena, alongside which possible exceptions must be carefully listed. In such a way generality, rather than universality, appears as the salient characteristic of scientific laws, while no ontological necessity is attributed them (again in keeping with Galileo's views).[3]

Universality and necessity, on the contrary, had been considered as the characteristic features of science (of *episteme*) by the classical tradition, and had remained substantially preserved in the view of science of the 'rationalist' representatives of modern philosophy. Therefore it is in a way surprising that these two features should be rehabilitated by Kant, who ascribed them to the two paradigmatic examples of 'science' he considers in the *Critique of Pure Reason* (and in the other works of his 'critical' period). These examples are mathematics and physics (this last being in effect Newtonian mathematical physics). Therefore he made a gigantic effort to explain how such a conquest could be attained by these sciences, and inscribed this effort in the 'epistemological dualism' he fully and explicitly accepted. Indeed, he distinguished 'phenomena' from 'things in themselves', and declared phenomena to be 'pure appearances'. In spite of this, phenomena are knowable, while things in themselves are not; and they are knowable because they are based on sensible intuitions (the 'sensible impressions') that are passively received by our sensible capabilities (be they those of the 'external' or of the 'internal' sense). The intellect, on the contrary, is not endowed with intuition: it is active, but its activity reduces to the capability of 'unifying' the content of sensible intuitions according to its own structural characteristics. Those characteristics are the 'pure concepts' or 'categories' that are no longer interpreted as the universal ontological features of being that the intellect abstracts from reality by means of its peculiar intuition, but simply as universal characteristics of human knowledge, unavoidably present in all knowledge since they are the very 'conditions of possibility' (or transcendental conditions) of knowledge as such. In such a way universality and necessity were recovered for any authentic knowledge, since they simply expressed the fact that unavoidably, and in all cases, we cannot know without using our intellectual a priori forms of knowing. The objects of knowledge are therefore 'constructed' by our intellect, but they are not 'produced' by it. Indeed Kant carefully distinguishes 'thinking' from 'knowing': thinking amounts to a pure combination of concepts, while knowing requires that these concepts be applied to actually present sensible intuitions. This is why Kant is concerned to distinguish his position from 'idealism' (that meanwhile had become the opposite of 'realism', as explained above). He qualifies his doctrine at the same time as 'empirical realism' and 'transcendental idealism'. The sense of this distinction resides in the issue of the 'dependence': realists maintained that the existence of the objects of our knowledge does not depend on our act of knowing them, idealists maintained that it does. According to Kant, the existence of these objects does not entirely depend on our act of knowledge, since the 'empirical' base is represented by sense intuitions that we do not produce, but are 'passively' received by us, while the construction of the objects follows the conditions imposed by our categories, and therefore depends on our intellectual a priori knowing capabilities.

These Kantian conceptions are so well known that we may dispense with illustrating them by means of quotations from his works. We want only to draw two conclusions from what we have said. First: while modern natural science has remained 'realist' in the ontological and epistemological sense (scientists admit that physical reality has an existence in itself, independent of our investigation, and that it is endowed with certain real characteristics, that are 'manifest' and can be known by us as they are), Kant gave a 'non-realist' interpretation of this science (what science knows is not reality in itself, or ontological features of reality, but a world of objects that are the organization of 'pure appearances' according to the transcendental conditions of our intellect). In spite of this he claimed to be, at least partially, a realist, in the sense that not everything in the objects of our knowledge 'depends' on us, because the sensible 'appearances' are only passively received by us. We find here a very peculiar form of epistemological realism, that undoubtedly prefigures Putnam's 'internal realism' (and Putnam himself duly recognizes this fact). We must also underline, however, that Kant's solution strictly depends on his unquestioning adherence to the 'epistemological dualism', which postulates the unknowability of things in themselves, and reduces the whole of knowledge to something 'internal' to the subject. This weak point was challenged by Kant's followers (already Jakobi had noted that 'without the thing in itself one cannot enter criticism [that is, Kant's 'critical' philosophy], but with the thing in itself one cannot remain in it'), and German 'transcendental idealism' has tried to eliminate this discrepancy between reality and thought. What remains to be seen is whether an 'internal realism' can be advocated without falling into epistemological dualism, a question that we will take up later.

For the moment we can note that Kant's 'phenomenalist' interpretation of science did not immediately achieve a large consensus. On the one hand, transcendental idealism, having eliminated phenomenalism altogether, considered natural science as a correct but still inadequate form of knowledge, that must be surpassed by a philosophical understanding of reality (the romantic 'Philosophy of Nature' that developed within the idealistic framework in practice amounted to a devaluation of science). On the other hand, the rich harvest of technological applications made possible by the rapidly increasing advances of the natural sciences easily convinced the general public that science is indeed an adequate knowledge of nature, and positivism gave to this spontaneous conviction a philosophical consecration, declaring science to be the unique form of adequate knowledge, and discrediting philosophy's rival pretension. This position obviously expressed a realist view of science.

This realist view began to enter a period of crisis in the second half of the nineteenth century. The construction of non-Euclidean geometries gradually discredited the role of mathematical intuition, showing that logically consistent geometrical theories can be constructed starting from intuitively mutually incompatible postulates. Even when set theory seemed to provide the bedrock foundation for the whole of mathematics, the discovery of the antinomies in this theory destroyed the confidence that we can intuitively know even such basic 'entities' as sets. As a consequence, mathematics came to be seen as a great family of logically interconnected hypothetico-deductive systems expressed in a formalized axiomatic way, whose legitimacy was not due to their capacity to describe the properties of 'mathematical objects', but simply to their internal logical consistency, or non-contradiction.

As to physics, the realist view of Newtonian mechanics was strongly reinforced during the first half of the nineteenth century. This was not only because of the impressive mathematical developments of Newtonian mechanics, but also because of the gradual appearance of a 'mathematical physics' that, concretely speaking, was nothing but an attempt to express, interpret, and explain the phenomena studied in the different branches of physics by means of the concepts, mathematical tools, and models provided by mechanics. The simultaneous formulation by different scientists of the principle of conservation of energy in 1847 seemed to offer a deep justification for the 'mechanistic world-view' that was advocated, for instance, by such scientifically outstanding and philosophically sensitive scholars as Helmholtz and Maxwell (since this energy was thought to be transformable, in the last analysis, into mechanical energy). Therefore the challenge for theoretical physics was seen to be to elaborate adequate 'mechanical models' for the two new branches of physics, that is, electromagnetism and thermodynamics.[4] This challenge, however, was doomed to failure, for no satisfactory mechanical model could be elaborated for the electromagnetic 'ether', and no satisfactory explanation of the second principle of thermodynamics could be provided within the framework of the kinetic theory of matter (in spite of the very ingenious efforts of several outstanding mathematical physicists on both problems). The reasons for these shortcomings soon appeared to be related to the fact that physics was seriously making its first steps into the realm of the unobservable. In this enterprise it made use of powerful idealizations, that were tacitly justified by two fundamental presuppositions, that is, that the laws and principles of mechanics have a true universality, and that their scope includes the microscopic as well as the macroscopic world.

Both these presuppositions were attacked by Ernst Mach, when he gave his diagnosis and therapy for this crisis of physics. His fundamental philosophical thesis was a form of radical empiricism, according to which only sense perceptions constitute knowledge. He did not deny a certain function to intellect. But he restricted it to the elaboration of general schemes that have no cognitive import, but only a pragmatic role, in the sense that they allow us to summarize sets of similar perceptions, to make useful predictions of future perceptual situations, and also to realize concrete applications. Intellectual constructions are simply conventions that can be abandoned and replaced whenever other conventions appear to be more useful. He added to this epistemological doctrine an ontological claim: unobservables are not simply unknowable, but also non-existent (indeed, he denied the existence of molecules). Therefore we must say that he expressed a clearly anti-realist view of science that was embedded in a more general philosophical anti-realism (since pure perceptions are not sufficient even for affirming the existence and knowability of common-sense realities such as cats, plants, or stones). More interesting is his criticism of the universality of mechanics, that had led to the fruitless efforts of constructing mechanical explanations of electromagnetic and thermodynamic phenomena. He pointed out that mechanics enjoyed the historical privilege of being the first exact natural science, which explains why scientists were spontaneously inclined to apply its concepts and principles in the study of newly discovered phenomena. Such chronological priority, however, cannot be taken as equivalent to any ontological, conceptual or logical primacy: therefore there is no justification for claiming that mechanics ought to provide the framework for interpreting the totality of physical phenomena. It must be noted that all the described events occurred long before the creation of relativity theory and quantum theory.[5] They only contributed to the deepening of the 'crisis' of classical mechanics, since they showed that many more concepts, laws, principles, and methodological presuppositions of this theory had to be fundamentally modified in order to satisfy the theoretical needs of the new physics.

Scientific realism (and anti-realism)

The rather lengthy story we have sketched above was necessary in order to propose a reasonable distinction between realism and anti-realism in general, on the one hand, and specifically scientific realism and anti-realism on the other hand. Indeed in many current discussions of realism and anti-realism that allegedly pertain to science, we simply find more or less elaborate variants of the positions about realism in general.[6] The problem of distinguishing general realism from scientific realism, however, is not simple. But we are simply interested, here, in understanding how it happened that a realist view of natural science could be generally held for a long while, and be almost suddenly abandoned towards the end of the nineteenth century. The reason why classical mechanics could receive a realist interpretation is that it remained in keeping with the spontaneous realism of common sense, to the extent that it appeared as a kind of 'prolongation' of common sense itself. Its concepts were certainly abstract, but at the same time could be seen as 'idealizations' of concretely observable physical bodies or events: a material point could be seen as the limit image of a grain of sand, a physical wave as the limit image of the waves in a pond of water, a rigid body as the limit image of an iron bar, a frictionless motion as the limit image of a perfect glass sphere moving on a perfectly horizontal ice surface, and so on. Though being concepts in a rigorous sense, they remained bound to observable physical objects or processes, they were visualizable (and this is why we have called them 'limit images'), and this spontaneously inclined people also to expect that other not explicitly encoded properties of the physical objects or processes from which the idealization had started should continue to be exemplified as well. Unfortunately this expectation was frustrated when models of the micro-world were put forth using the idealizations derived from the observed macro-world. The way out of this difficulty, in the spirit of classical physics, would have been to find new concepts obtained via idealization from the observation of the micro-objects, but they are unfortunately unobservable.

This is the frontier that separates contemporary physics from modern physics, since contemporary physics is essentially a physics of unobservable objects, and it is not by chance, as we have seen in our historical overview, that scientific realism began to be challenged when this frontier was encountered. Therefore we propose to characterize specifically the problem of scientific realism as the problem of the reality of the unobservables proposed by scientific theories. The suitability of this characterization is confirmed by the position defended by such influential scholars as van Fraassen: he accepts common-sense realism regarding the objects of everyday experience, since they are accessible to observation, and denies realism regarding the unobservable entities of natural science.

Can this position be defeated? A first step is to enlarge the concept of observation, based on the awareness that scientific observation is very different from pure sense perception, since it is essentially instrumental observation: we can say that we 'observe', for example, electrons through a suitable instrument like a Wilson chamber. This path is certainly fruitful, and scholars such as Shapere have developed it convincingly. The implicit assumption of this strategy, however, might be that observation (however enlarged) remains necessary condition for scientific realism. We maintain, instead, that it is a sufficient but not necessary condition, and that access to the reality of unobservables can also be obtained by means of the use of the intellect.

Realism and truth

At the beginning of this essay we noted that the problem of realism must not be mixed up with the problem of truth, since it is 'primarily' an ontological problem. This does not mean, however, that the ontological, the epistemological, and the semantical aspects of this problem, though distinct, should be separated. On the contrary, we shall see now that the reality of unobservables (ontological question) cannot be affirmed without resorting to truth and logic. Let us first remark that the notion of truth is intrinsically realist, since a judgement (intellectual level) or statement (linguistic level) can be said to be true if and only if 'it says of what is the case, that it is the case, and of what is not the case, that it is not the case'. This Aristotelian definition, that is already prefigured in Parmenides and almost literally formulated by Plato, not only corresponds to the common-sense conception of truth, but is implicitly or explicitly admitted by all philosophers in the whole history of Western philosophy. Differences (and notable differences) only regard the criteria of truth. Among such criteria intuition and logical consistency were soon identified as prominent. Intuition has been seen as the criterion for immediate truth, and logical consistency as the criterion for mediate or inferred truth. As to intuition, it was historically divided into sensible intuition and intellectual intuition (as we have already seen), and modern science predominantly emphasized sensible intuition or observation. As to inference, logic was created as the study of those intellectual (and linguistic) links among judgements that are truth-preserving, in the sense that if we apply them to true judgements (or statements) they necessarily lead to true judgements (or statements). This specific characteristic of logic is founded on the intentional identity of being and thought of which we have already spoken, and which can be summarized in the statement that thought cannot be thought of 'nothing', and therefore cannot help but being thought of being. This fact is reflected in the double formulation of the principle of non-contradiction: its logical formulation says that 'it is not possible to affirm and deny anything at the same time and under the same respect', and this because 'nothing can at the same time and under the same respect exist and not exist, or be such and such and not be such and such' (ontological formulation).

From these characteristics some consequences follow:

1. Whatever cannot be thought of (in the rigorous sense of being self-contradictory) cannot exist (logical contradiction entails ontological impossibility), and whatever can be thought of (in the sense of being

logically consistent) may exist (logical consistency entails ontological possibility).

2. If from an intuitively true judgement another judgement logically follows, the latter judgement is also true even when it is not intuitive.

3. Everything that exists must exist together with all things without which its existence would be contradictory.

This third statement is a sub-case of the second, once the ontological import of logic is accepted, since it says that the particular logical inference called *reductio ad absurdum* obliges us to admit the existence of even those unobservable realities without which features of the existing and observed realities would turn out to be contradictory. If we scrutinize the most rigorous arguments proposed by those metaphysicians who wanted to demonstrate the existence of meta-empirical entities, we can find that they more or less implicitly relied upon this statement.

Leibniz's principle of 'sufficient reason' ('everything existing must have sufficient reasons for its existence') constitutes an 'attenuation' of statement (3). He did not deny this statement (none who accepts the ontological import of logic can deny it), but admitted a flexibility in the determination of the conditions that make concrete reality intelligible (that is, non-contradictory). Therefore, in keeping with statement (3), he admitted that without the existence of God the existence of concrete things would be contradictory, but he thought that the logico-ontological dependence of the real world from God constitutes a chain that could have been different, since other chains could have provided a logically consistent set of 'sufficient reasons' for the existence of the world. The actual world does not result from logical necessity, but from free divine decision, and is therefore 'the best of all possible worlds', where 'the best' expresses the choice of divine wisdom that must not be confused with logical necessity. In this sense the world is contingent. It was Hegel who embarked on the ambitious enterprise of endowing the whole of reality with logical necessity (and in his case it may be right to say that he claimed to identify knowledge of reality with a 'God's Eye point of view').

What has all this to do with scientific realism? Scientific theories try to satisfy the Leibnizian principle of sufficient reason: indeed they try to offer, by introducing a certain number of unobservable entities endowed with precise properties, a global picture in which the observable objects of their domain can be logically explained, that is, can be shown to be logically connected with these entities in such a way that sufficient reasons for their observable features can be offered. But, one might say, how can one be sure that the unobservable entities 'really exist', since any theory only provides sufficient reasons, and we all know that – all theories being 'underdetermined' with respect to observable data – several different theories could be proposed capable of providing sufficient reasons for the same data? In this question several aspects are unduly confused, and we shall try to disentangle them:

1.   The underdetermination of theories is a strictly epistemological fact, that in the last analysis reduces to the well-known logical fact that the truth of the conclusion of a correct argument is not a sufficient condition for granting the truth of the premises. This 'granting' actually means that, in this particular case, logic does not provide a fully reliable tool for knowing whether the premises are true or not. But this does not at all affect the fact that the premises are in themselves, and necessarily, either true or false, since this is an intrinsic semantical property that is independent of any epistemological condition. For example, if the only observational evidence available to me is that my wallet is no longer in my pocket, I can formulate different hypotheses that can logically explain this fact: that I have left it in my office, in the restaurant, in the supermarket, that I lost it in the bus, and so on, and I do not know which of these different hypotheses is true. However it is obvious that at most one of them is true, and perhaps none (for example, my wallet has been stolen by a pickpocket). In conclusion: underdetermination of theories only entails the epistemic situation that we never attain absolute certainty about their truth.

2.   The nature of truth, expressed in its semantic definition already mentioned, contains an ineliminable ontological component: a statement is true if and only if the 'state of affairs' it describes really exists. If I hold that the statement 'there is a copy of Kant's *Critique of Pure Reason* in the library of my department' is true, I must necessarily hold that this book really exists in this library, independently of the fact that I am seeing it at the moment of uttering my statement. Coming to scientific theories, we must say that if one holds that a certain theory is true, one cannot consistently avoid holding that the 'states of affairs' it describes really exist, independently of the fact that such states of affairs involve unobservable entities. As a consequence, the only correct way to deny the existence of the unobservables introduced in a scientific theory is to say that the theory is false. If we are in the position of holding (even without absolute certainty) that the theory is true 'beyond any reasonable doubt', we must also

hold 'beyond any reasonable doubt' that the unobservables it introduces really do exist.

Is it possible to escape these conclusions? Certainly it is, but at a very high
cost. A first move can consist in denying that scientific theories can be
qualified as either true or false. There is a grain of reasonableness in this
position, since truth and falsity have been directly defined for single
statements, and not for sets or systems of statements such as whole
theories. There is, however, the possibility of 'broadening' this definition
to include in an 'analogical' sense the notion of true and false theories. The
ground for denying truth or falsity for theories, however, usually consists
in a much less justifiable philosophical tenet: radical empiricism, which we
have seen to play its devastating role already with Mach. This tenet is
totally dogmatic, since it reduces the cognitive capability of humans to
sense perception, contrary to common-sense evidence and also to the
evolutionary interpretation of the characteristics of humankind. In fact, we
(rightly) believe that we know that Caesar crossed the Rubicon, that he was
killed in the Roman Senate, that Napoleon was defeated at Waterloo, and
so on, without having any possible observational access to these past
events, but simply because we confidently rely on such instruments of
historical knowledge as documents, photographs, and records of facts, that
induce us to hold 'beyond any reasonable doubt' that these individuals
really existed, and that true stories are told about certain past events.
Similarly we claim to know of the existence of numberless cities and
regions of our planet we have never visited. In short, though this
knowledge has originated in the 'observation' made by a few single
individuals, it can become knowledge for many other individuals only
thanks to a very complex cognitive role of human intellect (intersubjective
knowledge is impossible within pure sense perception or observation). One
might wonder why such total confidence is placed in observation. The
most plausible answer is that direct observation is endowed with certainty.
A slight acquaintance with the history of philosophy, however, abundantly
shows serious criticisms of sense knowledge and, if the opponents of the
reliability of the intellect can mention such stock examples as phlogiston,
opponents of the reliability of observation can mention such equally stock
examples as the stick that appears bent in water, or the earth that appears
motionless at the centre of the universe. The real situation is that human
knowledge is constituted by an interplay of both sense and reason, and that
the cooperation of both enables man to attain truth about reality, in its
observable and unobservable dimensions.

Objects and properties

A more subtle objection that can be raised against the existence of unobservable entities is that all that scientific theories can offer is the determination of certain properties of the world (usually expressed by mathematical relations and equations), but not of the objects that allegedly possess such properties. Therefore, while a certain realism of properties could be accepted, a realism of entities should be rejected (Poincaré, for example, already admitted a limited 'realism' of science in the sense that only the relations between the phenomena discovered by science are to be taken as real, not the objects that theories claim enter into such relations). In spite of seeming very modern, this issue is actually very old: in the vocabulary of classical ontology distinguishing 'substance' from 'accidents', medieval philosophers had already noted that, while a substance is not identical with its accidents, we can only know a substance through its accidents. A substance, as we know, is an individual entity existing in itself, while accidents are the qualities and relations that can only exist in a substance and not in themselves. If we actually know an individual substance – for example, a given dog – even at the level of pure observation, what we know are several accidents (in this case, sensible qualities), but certainly not all of its qualities (for example, its being the most cherished living thing of an old lady): nevertheless we rightly say that we know this dog, since these qualities are 'its' qualities. In other words, to know an individual amounts to knowing 'what it is like', and this means knowing certain features of it that are, so to speak, intrinsic and stable, such that they enable us to identify it, and also to re-identify it in different conditions. As Strawson[7] has rightly pointed out, in order to identify an individual object we need to use a 'sortal', that is, a term indicating 'what sort of object it is', and this sortal unavoidably contains in its meaning an organized system of properties that are concretely realized in the individual. Even theologians, when they try to characterize the particular individual entity that is God, cannot help but doing so by means of certain properties (such as omniscience, omnipotence, eternity, immutability, creative power, and so on). In conclusion, since accidents (or qualities) cannot exist in themselves, but are always qualities of something, it is impossible to maintain that we can know qualities that are qualities of nothing: we know something through its qualities.[8]

If all that we have said is true of knowledge in general, it must be true also of science in particular. We can easily verify this by reconsidering the breakthrough which constituted the core of the 'Galilean revolution'. A realizable knowledge of the 'natural substances' must renounce the pretension of grasping their intimate essence, and be content with knowing some of their 'affections' or 'accidents'. Therefore, natural science was considered to be a knowledge of natural substances, attained through the knowledge of a selected set of their accidents. This methodological approach has become characteristic of every empirical science in the modern sense of the concept. Leaving aside the nowadays old-fashioned terminology of 'substance', 'accidents', 'affections' and the like, we shall say that every empirical science presents itself as knowledge of reality, not in general, but from a specific 'point of view', that amounts to considering only a selected set of attributes of things. This set of attributes delimits the domain of objects of the given science, in the sense that only those things that are endowed with all these attributes can become the 'object of study' of the said science. Mechanics, for example, selects as attributes of things with which it will be concerned only mass, location in space, location in time, and force. Therefore, on the one hand, concretely existing and perceivable things such as a toothache, or a symphony, cannot be objects of mechanics because they fail to possess at least one of mechanics' specific attributes (for example, mass). On the other hand, one single thing may become the object of several different sciences, according to the different points of view from which it is investigated. A cat, for example, is an object of mechanics when we try to explain how it can fall from a window and yet reach the ground on its four paws without infringing any law or principle of the mechanics of falling bodies, but it is an object of chemistry if we study the composition of its hair, an object of animal psychology when we study how certain conditioned reflexes come about in its behaviour, an object of economics when we want to explain why it is sold at a very high price on the market, and so on.

Any mature science determines its domain of objects when it is able to characterize its selected attributes by means of clearly defined concepts that are also mutually correlated. These concepts are linguistically expressed using certain predicates that make up the 'technical vocabulary' of a given science, and their correlations (that also amount to a 'contextual definition' of their specific technical meaning) are expressed in certain fundamental statements, often called laws or principles. In the case of physics these predicates are magnitudes and these statements are mathematical expressions (in the majority of the cases, equations or 'disequations'), but this additional condition is not required for other sciences. To sum up: every (empirical) science is concerned with certain attributes of things (ontological level), that are abstracted and idealized in certain concepts (intellectual level), and expressed by means of certain predicates (linguistic level). Since every science aims at knowing reality, though only from its specific point of view, it must equip itself with certain tools of reference that enable the scientist to ascertain whether or not a given 'thing' really possesses a certain attribute in the way described by a statement of his science. Owing to the fact that scientific knowledge seeks to be intersubjective, such tools are provided by standardized operational procedures that are neither linguistic nor mental, but are as much concrete as things, and which enable one to know whether a proposed statement is immediately true of that thing, or not. This entails that at least certain predicates of an empirical science must be also operationally defined, in the sense that they are related to certain precise operational procedures of referentiality. It is thanks to these procedures that a scientific concrete object can be 'cut out' of a concrete thing. We will call such operationally defined predicates 'basic predicates', and say that they permit us to construct scientific objects, not in the sense of creating them, but in the sense of determining them as particular aspects of reality. For example, the predicate 'mass' of classical mechanics is equipped with the operational procedure of putting things on a balance in order to check the truth of the statement that a certain thing has a mass of x grams (within the limits of precision of the balance employed). The states of affairs resulting from such operational procedures are the data of a particular science, that correspond to the immediately true statements it must accept according to its own constitutive conditions.

The discourse developed thus far holds for every empirical science, but natural sciences are characterized by the fact that they do not study single entities or processes, but general properties of indefinitely large classes of things and processes, and this entails that the actual contents of their investigation are those ideal objects that result from the combination of the specific concepts and conceptual correlations we have mentioned above. To make things easier we will bracket for a moment the mental level and consider directly the linguistic level. From this point of view, a scientific object becomes a structured set of predicates, and all objects that are introduced by a scientific theory are of this kind. We must therefore say that scientific objects (in this new sense) are abstract objects, such as the previously considered examples of material point, rigid body, elastic recoil, frictionless motion, perfect gas, and so on. We must now remember that any abstract object (that is, any well-defined concept) is totally and adequately characterized by the finite collection of properties it encodes. In this sense it is not appropriate to say that it has its properties, but that it *is* its properties. If we want to call it an 'entity' (and this is not arbitrary since it has a mental existence, can be identified, and receive a proper name), we can say that knowing the properties amounts to knowing the entity in the full sense, that is, without any residue.

We must not forget, however, that the purpose of natural science is to know nature, and this entails that at least some of the abstract concepts of a natural science must also be exemplified by concrete things. This condition is fulfilled by the fact that, thanks to the operational tools, at least some logical consequences derived from the admission of abstract entities must be exemplified by experimental tests, that is, recognized as immediately true. Since, however, we have already seen that the truth of the consequences may also lead us to admit (even without absolute certainty) the truth of the premises, we can say that, if such a truth is established beyond any reasonable doubt, we must also admit that the abstract entities are exemplified, that is, that they have their (unobservable) referents, that they are not pure concepts or linguistic constructions. In other words, there exist electrons and elementary particles, and not only the concepts of 'electron' and 'elementary particle'.

Let us come to a final and important point. In the above reflections we have stressed that the admission of unobservable entities is legitimate when it allows for a correct account (and, we must add here, also for successful predictions) of operationally testable events. Such an 'account' is usually interpreted as a logical justification, but this is only a consequence of the 'statement view of theories' that is in turn a consequence of the 'linguistic turn' that has also affected in particular the philosophy of science. One must recognize, however, that all this is simply a logico-linguistic translation of something which, for common sense and also for scientists, is a causal explanation. The occurring of certain concrete states of affairs cannot be seriously conceived as being produced by conceptual or logical operations. Concepts and statements can only 'produce' concepts and statements by means of intellectual operations, they cannot produce concrete things or events. From this point of view, the affirmation that, outside a realist conception of science, the widespread success of scientific predictions would be a miracle is far from being so trivial as certain philosophers of science maintain. On the contrary, the constant and even overwhelming development of technology is a gigantic confirmation of the realist import of the scientific theories that technology applies, since it does not apply them to conceptual debates, but to the real world.

Realism and perspectivism

What we have presented allows us to see that realism is fully compatible with perspectivism, since the fact that we consider reality simply from a certain point of view (or 'perspective') does not eliminate the fact that, in such a way, we are considering or singling out certain real aspects of reality, aspects that 'belong' to it, even if they could be uncovered only by resorting to a particular perspective. For example, conic sections are 'really' contained in a cone, though we can bring them to light only by 'cutting' the cone by means of a suitably oriented plane. If the plane is parallel to the base we obtain a circle, while we obtain an ellipse or a hyperbola if the plane has a different inclination. These geometrical figures can be obtained by cutting an ideal geometric cone by means of an ideal geometric plane, but if we cut a material cone by means of a materially realized plane, we can actually observe geometrical shapes that exemplify the conic sections. That they are 'real attributes' of the cone immediately results from the fact that we cannot obtain them by 'cutting', let us say, a pyramid or a cube. Moreover, these figures may be described and treated in several ways and according to several languages: for example, we find in the textbooks a 'projective theory of conic sections', as well as an 'analytic theory of conic sections', in which their properties are described and investigated from the 'point of view', or 'within the perspective' of projective geometry or analytic geometry respectively. The broadening of the mathematical investigation can lead to the creation of specialized technical expressions that are only indirectly reminiscent of the original situation, for instance when we speak of 'elliptical equations' in analysis, or of 'elliptical geometry' for characterizing a particular kind of non-Euclidean geometry. At the same time, other concrete objects may be found that exemplify the concepts of these conic sections, like the planetary orbits, that turned out to be elliptical, and not circular, as they had been believed to be at first.

This discourse is true in general: every property of a thing really exists in the thing, on the one hand, but can be known only within a suitable perspective. For example, if I observe a rose under 'normal' sunlight conditions and I find that it is red, while the grass around it is green, I must say that the property red is real in the rose, and the property green is real in the grass, since I cannot help but see them this way, and I cannot see the rose as green and the grass as red. All this, however, does not simply depend on the intrinsic properties of the rose and the grass, but also on suitable lighting conditions and on the specific constitution of my visual apparatus. If the rose or the grass are observed under some peculiar artificial light, or by an eye affected by certain disturbances (for example, by colour-blindness), the perceptual colour of these same objects will really be different. It is only for the needs of intersubjective communication that we say that the real colour of concrete material objects is that which is perceived by a 'normal' eye under 'normal' light conditions. If we 'observe' the rose or the grass by means of a standardized physical apparatus, such as a photoelectric cell, we might believe that we have access to their 'real' colour. But this is a misunderstanding: such an apparatus will tell us that electromagnetic waves of a certain predominant wavelength have been emitted by the surface of these physical bodies. But this is by no means a 'perceptual colour', since it can only be thought of from this point of view, and even a blind person can say that the rose is red, without any possibility of perceiving it as red. This possibility of thinking and saying is based only on the translation of the perceptual vocabulary into a physical-theoretical vocabulary made possible by those who have 'linked' the 'normal' perceptual red with the 'standardized' physical red, for they had direct access to both of them.

Since, as we have seen, realism is fully compatible with perspectivism, we can conclude that realism does not imply a 'God's Eye point of view'. If we think of God as omniscient, we may say that he is able to know reality from all possible points of view, and not from no point of view, or from nowhere, since his infinite cognitive capability allows him to know all aspects of reality from an adequate point of view. Human knowledge is finite, and therefore we can only know attributes of reality that are accessible to man from a limited number of points of view: those that are constituted by his sense organs, and by his intellectual power (which, in particular, enables him to construct instruments for broadening the possibilities of his 'observation'). We can also express these conclusions by saying that genuine knowledge need not be total knowledge (this might be the privilege of divine knowledge), and also that truth (specifically considered as a property of judgements, statements, theories, that is, of ways of expressing knowledge) is always relative, that is, on the one hand, relative to the referents of the discourse and, on the other hand, relative to the cognitive tools adopted. Absolute truth (in the sense of being independent both of referents and of cognitive capabilities) is not a privilege even of God.

External and internal realism

Our reflections enable us to maintain that realism (including scientific realism) must be considered at the same time (but under different respects) as 'external' and 'internal'. These adjectives are rather pictorial, and have a topographical shade of meaning that must be clarified in a conceptual way. This is not easy to determine, and the same Putnam who insisted on this distinction, was actually able only to mention a list of features that should characterize these two philosophical perspectives, without really 'defining' them. We can venture, however, to give a rather precise characterization by taking knowledge as reference point, and calling 'external' that form of realism that maintains two theses:

1. That reality consists of ontologically existing things with all the attributes that characterize each of them, and this existence does not depend on our knowledge of them (that is, this knowledge is not a condition for their existence).

2. This reality is the content of our knowledge, in the sense that we could not call knowledge in a proper sense any mental or linguistic construction whose referents would not be really existing entities or properties.

By way of opposition we can call 'internal' that form of realism that maintains:

1. That reality depends on knowledge in the sense that – as Putnam says – '"objects" do not exist independently of conceptual schemes'.[9]

2. That referents of our knowledge are objects that are constructed by our cognitive capabilities themselves.

There is a sense according to which internalism is irrefutable, but this is a very trivial sense: reality of which we can speak, affirm the existence, know properties, and so on, is strictly reality that is included within our knowledge. In fact, it is obvious that 'speaking of', 'affirming', and so on are ways of expressing our knowledge. Therefore, claiming that we cannot know reality 'outside' knowledge, or 'independently of' knowledge is an epistemological truism. Indeed, not even the most radical 'external' realist would deny that things and properties, though existing independently of any knowledge (ontological level) cannot be known, and their reality cannot be affirmed, outside knowledge. Therefore the correct position seems to be expressible in this intentionally pictorial way: things exist 'outside' the mind (or outside theories), but can be known only when they are 'internalized' within the mind (or within theories). Therefore, Putnam is right to say that '*what objects does the world consist of*? is a question that it only makes sense to ask *within* a theory or description',[10] and this is simply because in order to answer this question we must know how the world is, and this knowledge is necessarily expressed in a theory or description. However, we can maintain that this question is answered correctly only if we are convinced that the description says how the world really is.

Is there a possibility of satisfying this requirement within an internalist perspective? There is one, which consists in maintaining that things are nothing but intellectual constructions. But this move would reduce internal realism to idealism, something Putnam does not want. For this reason he explicitly subscribes to Kant's solution, admitting that 'there are experiential inputs to knowledge; knowledge is not a story with no constraints except *internal* coherence, but [internalism] does deny that there are any inputs *which are not themselves to some extent shaped by our concepts*'.[11] These inputs, as stated a few lines above, are 'the "objective" factor in experience, the factor independent of our will'. In conclusion, the nail on which 'realism' is hung is the 'passivity of sensation' already claimed by Kant and, exactly as in Kant, depends on the dogmatically presupposed epistemological dualism we have already discussed above.

The position we are advocating, and which we preferred to call 'metaphysical realism' in the richer sense already explained, simultaneously affirms the ontological existence of things and properties, that are open to knowledge since they are in part sensible and in part intelligible. The suffix '-ble' usually indicates a potentiality (as in thinka*ble*, feasi*ble*, practica*ble*, prefera*ble*, and so on), and in our case indicates that certain properties can give rise to sensations or concepts if they 'stimulate', so to speak, either certain well-determined sense capabilities or else our intellect. This stimulation, however, must not be confused with a purely physical action (therefore, the criticism developed by Putnam and other authors against a 'causal' theory of sensation or perception, as well as of the 'similarity' theory of the same, is well grounded). This stimulation is actually a simple 'intentional presence', that 'internalizes' the sensible and intelligible properties not in an ontological sense (they remain properties of things, and do not become properties of the eye or of the mind), but so as to permit the actualization of the cognitive capabilities of sense and intellect. These cognitive capabilities are in part inborn, and in part modified by the accumulation of past knowledge and by several cultural factors, so that the 'conceptual schemes' of which Putnam speaks are certainly part of this interplay between reality and mind, but they simply help men see reality from certain points of view or perspectives, that can often lead to the discovery of previously unknown attributes of reality, but are not automatically granted to 'match' ontologically existing properties of reality. It is certainly significant that Putnam, and many other scholars discussing the issue of realism, do not even mention this fundamental notion of intentionality.

Another notion that is inadequately discussed by the same authors is that of 'reference', and this is not surprising for reference is intimately connected with intentionality. Referents are those entities that do not belong to the same ontological level as the discourse that is about them, and for this reason they must be attained by other means than pure thinking or pure speaking. We have already explained that every empirical science must equip itself with operational tools of reference, and this already stresses that it is through a certain 'doing' that we try to meet the referents, a doing that is certainly 'guided' by meaning and theoretical constructions, but whose outcome is by no means predictable or predetermined by any theoretical or linguistic manoeuvre. Precisely these criteria of reference determine the ontological status of the entities about which a theoretical discourse tells its story. A referent is an entity that exemplifies a concept or a theoretical construction: such concepts or constructions, as we have seen, constitute the abstract objects of a theory or description and as such have a purely mental existence, but the referents or concrete objects of the theory or description must be detected by extratheoretical means. For example, not only must the concept of a rigid body be exemplified (within an acceptable approximation), by certain material bodies, but also the affirmation that Hector is a Trojan warrior in the *Iliad* can be said to be true only if, by a particular operation (reading the *Iliad*) we find that this 'literary entity' really possesses (in the *Iliad*) the property of being a Trojan warrior.[12] Of course, the 'linguistic turn' that has in particular led to an almost complete 'contextualization' of meaning, has almost fatally produced a conflation of meaning and reference. This, however, is a shortcoming of this philosophical approach. It must be overcome if we want to discuss the issue of realism in all its complexity, and, in particular, recover the autonomous sense of referentiality. This means that, as Putnam says, we must give up the 'similitude theory of reference' (which, by the way, was not advocated by classical epistemology), but not in order to accept a contextual theory of reference, that is equally unsatisfactory. What we propose is an operational theory of reference that enables our

Evandro Agazzi

discourses to be about a reality that is not produced by the discourse, but only intentionally present to it when it is concretely met through concrete operations.

Notes

1 Putnam (1981, p. 49).
2 For a detailed discussion of Galileo's position, see Agazzi (1994), which provides the necessary quotations from his writings, that were published in Galileo's *Opere*.
3 These general methodological principles are clearly summarized, for instance, in the *Scholium Generale* of Newton's *Principia* (Newton 1687) and, in a more elaborated form, in Question 31 of Book 3 of his *Opticks* (Newton 1704).
4 For example, even J. C. Maxwell, in the last pages of his *Treatise on Electricity and Magnetism*, indicated as a task for future generations that of finding a mechanical description of the electromagnetic field for which he had offered his famous equations. For details on this issue see, for example, my large introduction to my Italian translation of Maxwell's work (Maxwell 1972), as well as Agazzi (1975).
5 Mach's work on the historical development of mechanics, which substantiates his views on this issue, was actually published in 1883 (cf. Mach 1883).
6 This is the case, for example, even with a famous philosopher of science such as Popper.
7 Strawson (1959).
8 Classical epistemology was perfectly aware of this fact, expressed in a very eloquent principle: *talia sunt subiecta qualia permittuntur a predicatis suis* (subjects are such as they are permitted to be by their predicates), which means, ontologically, that entities are such as their properties allow them to be. In other words, properties are 'constitutive' of entities.
9 Putnam, op. cit., p. 52.
10 Op. cit., p. 49.
11 Op. cit., p. 54.
12 For a detailed discussion of this issue see Agazzi (1997).

References

Agazzi, E. (1975), 'De la théorie électroélastique à la théorie électromagnétique du champ', *Dialectica*, 29/2–3, pp. 105–26.
Agazzi, E. (1994), 'Was Galileo a Realist?', *Physis*, 31/1, pp. 273–96.
Agazzi, E. (1997), 'On the Criteria for Establishing the Ontological Status of Different Entities', in *Realism and Quantum Physics*, E. Agazzi (ed.), Amsterdam/Atlanta, Rodopi, pp. 40–73.
Galileo, G. (1929–39), *Opere*, Edizione Nazionale, Firenze, Barbera, 20 vols.
Mach, E. (1883), *Die Mechanik in ihrer Entwicklung historisch-kritisch dargestellt*, Wien.
Maxwell, J. C. (1972), *Trattato di elettricità e magnetismo*, Introduction, notes and translation by E. Agazzi, Torino, UTET, 2 vols.

Newton, I. (1687), *Philosophiae Naturalis Principia Mathematica*, London.

Newton, I. (1704), *Opticks*, London.

Putnam, H. (1981), 'Two Philosophical Perspectives', in H. Putnam, *Reason, Truth and History*, Cambridge, Cambridge University Press, pp. 49–74.

Strawson, P. F. (1959), *Individuals. An Essay in Descriptive Metaphysics*, London, Methuen.

Chapter Four

Realism, Method, and Truth[1]

Howard Sankey

Rational scientific inquiry is governed by the rules of scientific method. Adherence to the rules of scientific method warrants the rational acceptance of experimental results and scientific theory. Scientists who accept results or theories licensed by the rules of method do so on a rational basis. Thus, rational justification in science is closely connected with scientific method.

But while it is evident that there is a close relation between method and rational justification, substantive questions remain about the relation between method and truth. For example, are scientists whom method licenses in accepting a theory or experimental result thereby licensed in accepting the theory or result as true? Does use of scientific method lead scientists to discover the truth about the world? Questions such as these are questions about the truth-conduciveness of method. While they relate directly to the epistemic status of method, they bear indirectly on the nature of rational justification. For if use of method conduces to truth, then, given the relation between method and justification, the warrant provided by method is warrant with respect to truth.

Questions about the relation between method and truth divide scientific realism from anti-realism in the philosophy of science. On the one side, scientific realists take the aim of science to be discovery of the truth about the world. Realists defend the view that employment of the methods of science promotes the aim of truth. On the other side, anti-realists in the philosophy of science deny the connection which realists see between method and truth. Anti-realists typically agree that method underwrites the rationality of science. Some anti-realists deny that there are good grounds for taking use of method to lead to the realist aim of truth. Other anti-realists object to the realist conception of truth, and deny that method promotes truth in the sense intended by realists.

In the present context, the key question that divides scientific realism from anti-realism about science is whether employment of method advances the realist aim of truth. This is a question about whether a proposed means for the achievement of a given end is in fact a means conducive to that end. More specifically, it is the question of whether good grounds may be given for taking the methods of science to promote the realist aim of truth.

My aim in this chapter is to defend the realist response to this question by arguing that there are strong abductive grounds for taking the methods of science to be truth-conducive. Before I turn to that task, let me first address the relation between method and rational justification in somewhat greater detail.

Scientific method and rational justification in science

In this chapter, I assume a traditional view of the relation between scientific method and rational justification in science. On such a view, there is a close connection between scientific method and the rational acceptance of scientific theories and experimental results. In particular, compliance by a scientist with the rules of scientific method rationally justifies the scientist's acceptance of a theory or result. A scientist whose acceptance of a theory or result fails to comply with the rules of method thereby fails to accept the theory or result on a rational basis.

However, while I assume a traditional view of the relation between method and rational justification, I do not assume a traditional view of the nature of method itself. The traditional view of method is a monistic view, according to which there is a single, historically invariant method, the use of which is the characteristic feature that distinguishes science from non-science. By contrast with the traditional monistic view, I adopt a position of methodological pluralism according to which there is a set of methodological rules which scientists employ in the evaluation of alternative theories and the acceptance of results. These rules are subject to variation in the history of science, and different rules may be employed in different fields of science. Given the plurality of rules, scientists may diverge in the rules they employ, with the result that there may be rational disagreement among scientists on matters of fact and choice of theory. On such a pluralist view of science, while no single method is characteristic of science, the sciences are generally characterized by possession of a set of methodological rules which inform the factual and theoretical decisions of scientists.[2]

Much remains to be said about the relation between method and rational justification. However, for present purposes, I will assume that the relation between method and rational justification is straightforward. The purpose of this paper is to examine the relation between method and truth. Even if we assume that compliance with the rules of method justifies acceptance of a theory or result, the question remains of whether the theory or result is to be accepted as true. There is an epistemic gap between method and truth. My aim is to bridge this gap.

The realist conception of truth

It is often said that the conception of truth best suited to realism is a correspondence conception of truth. On such a conception, truth is a property which a statement has in virtue of a relation of correspondence that holds between the statement and the way the world is. A statement is true just in case what the statement claims to be the case is in fact the case. The relation of correspondence is, therefore, a relation between language and reality. For it is a relation between a statement couched in a language and an extralinguistic state of affairs that obtains in reality.

Since a statement is true just in case the state of affairs to which it corresponds obtains, the correspondence conception satisfies the equivalence condition specified by Tarski's T-scheme:

(T) 'P' is true iff P.

While the T-scheme is not a definition of truth, it provides a minimal condition of adequacy that must be satisfied by any account of truth. However truth is conceived, the truth-predicate must behave in accordance with the T-scheme. Rather than a definition, the T-scheme is a schema on the basis of which metalinguistic statements of truth-conditions may be formulated for sentences of an object-language.[3] For example, replacing 'P' in (T) by 'Electrons have negative charge' yields as statement of the truth-conditions of 'Electrons have negative charge' the T-sentence:

(E) 'Electrons have negative charge' is true iff electrons have negative charge.

Statements such as this assert the material equivalence of sentences that predicate truth and the sentences of which truth is predicated. The T-scheme thereby specifies a correlation between the truth of statements and the states of affairs that statements report. For it stipulates that, for any sentence 'P', 'P' is true just in case a given state of affairs obtains, namely, the state of affairs that P.

But to capture the thought behind the realist conception of truth, it is not enough to say that a statement is true just in case a given state of affairs obtains. That suggests that the relation that obtains between the truth of a statement and the state of affairs that it reports might be a mere accidental correlation. But it is no accident that a statement that reports a state of affairs is true if, and only if, the state of affairs it reports does in fact obtain. For it is precisely the fact that the state of affairs obtains that makes the statement true. It is because electrons in fact have negative charge that the statement that electrons have negative charge is true.

Yet even if we insist that statements be made true by extralinguistic states of affairs this does not suffice for a realist conception of truth. More must be said about the nature of the extralinguistic reality that makes statements true. There are any number of non-realist positions for which statements are made true by extralinguistic states of affairs. The idealist who takes the world to be ideas in the mind of God may say that statements are made true by ideas in the mind of God. The phenomenalist who identifies reality with the permanent possibility of experience may say that statements are made true by the permanent possibility of experience. But the realist can accept neither the idealist nor the phenomenalist scenario. For it is a defining feature of realism that the reality investigated by science is an objective reality that is neither constituted nor determined by thought or experience.

To rule out such mentalistic scenarios, the realist must insist that what makes statements true or false are states of affairs whose existence is in no way dependent on the mental. To qualify as a realist conception of truth, the correspondence theory of truth must be supplemented with the metaphysical realist assumption of a mind-independent reality. On the realist conception of truth that results, truth consists in correspondence between a linguistically formulated statement of fact and an extralinguistic state of affairs, where the state of affairs that makes a statement true is a mind-independent state of affairs. If it is true that electrons have negative charge, then this is due to the fact that, independently of anything we think about the matter, there are electrons, and they do indeed have negative charge.[4]

The non-epistemic nature of realist truth

The realist conception of truth is a non-epistemic conception of truth, which enforces a sharp divide between truth and rational justification. One may rationally believe a proposition that is false, just as there may fail to be rational grounds to believe a proposition that is in fact true. Far from being an absurd consequence of realism, as some may think,[5] the non-epistemic character of truth crucially underlies the central epistemological claim of scientific realism, namely that there is an epistemic gap between method and truth which is best spanned by means of realist resources.

It is important to distinguish between two different senses in which the realist conception of truth is a non-epistemic conception of truth. The first sense is a metaphysical sense, which derives from the mind-independence of the states of affairs that make statements true. The second sense is a conceptual one, which is due to the lack of a conceptual relation between truth and rational justification.

In the first sense, the non-epistemic nature of realist truth derives specifically from the mind-independent status of the truth-makers. The point turns on the ontological independence of thought and reality, rather than on any epistemic aspect of the relation between thought and reality. For the truth of claims about the world is solely determined by the existence of states of affairs which obtain independently of human thought or experience. Hence, the belief that a given state of affairs obtains does not itself – that is, *qua* belief – have any effect on the truth or falsity of that belief. The state of affairs may obtain, or fail to obtain, whether or not anyone believes that it does. This remains the case regardless of how well justified the belief may be. Thus, given the mind-independence of the truth-makers, it is entirely possible for rationally justified beliefs about the world to be false. Indeed, given such mind-independence, the entirety of such beliefs might be false.

The second source of the non-epistemic character of realist truth is the lack of a conceptual relation between the concept of truth and concepts of epistemic justification. On the realist conception of truth, truth is a relation of correspondence that obtains between statements and mind-independent states of affairs that obtain in the world. A statement is true just in case an appropriate state of affairs obtains. Thus, truth depends solely on the way the world is, whether or not the world is rationally believed to be that way. As such, no epistemic condition enters into the realist conception of truth.

More specifically, to be true in the realist sense a statement need not fulfil any epistemic condition, such as evidential support or the satisfaction of methodological rules. It need only reflect the way the world is. Nor is any epistemic concept built into the realist conception of truth, since formulation of the latter makes no use of concepts of rational justification or methodology. Hence, a statement may be epistemically well justified, in the sense of satisfying relevant methodological rules, and yet fail to be true. Indeed, a statement may be ideally justified and not be true, since no entailment from epistemic justification to truth is licensed by the realist conception of truth.

Both of the foregoing senses in which realist truth is non-epistemic reflect important principles of realism. The first reflects the fundamental metaphysical tenet of realism that the world investigated by science is an objective reality that lies beyond the control (though not the reach) of human thought. The second stems from the realist view that the truth of a claim about the world consists in correspondence with such an objective reality, rather than in satisfaction of criteria of epistemic evaluation.

In light of the non-epistemic nature of realist truth, the basis of the epistemic gap between method and truth is now apparent. It is not just that it is an intelligible question whether a belief warranted by the rules of method is to be accepted as true. The point is deeper than that. Because truth depends on a mind-independent reality, and is not defined in terms of epistemic criteria, a theory might fully satisfy relevant criteria and still be false. Conversely, a theory or claim about the world might be true even though it fails to fully satisfy applicable rules of method. Given the non-epistemic nature of truth, there is no logical relation between method and truth. The question must inevitably remain open whether the methods employed in science really do lead to truth.

Two anti-realist strategies

I will now consider two of the principal anti-realist strategies for dealing with the relation between method and truth. Since my aim is to provide a realist bridge between method and truth, I will not attempt a detailed examination of anti-realism here. Still, to understand the realist project, it is important to contrast it with alternative approaches to the problem.

The two strategies to be considered here represent opposing anti-realist tendencies. They are the Scylla and Charybdis between which the realist must steer a course. The first strategy is that of the 'internal realism' proposed by Hilary Putnam and Brian Ellis. The internal realist strategy is to bridge the epistemic gap by defining truth in terms of method, which creates an analytic relation between method and truth.[6] The second strategy, found in Bas van Fraassen and Larry Laudan, is one that I refer to as 'scientific scepticism'. The sceptical strategy treats the gap between method and truth as one that cannot be bridged. It denies that satisfaction of method licenses rational belief in truth. Instead of truth, scientific sceptics offer alternative epistemic aims which they take to be achievable using the methods of science.

While detailed critique of either form of anti-realism lies beyond the scope of this chapter, it is worthwhile situating the two positions with respect to realism. By contrast with realism, the internalist denies that there is a gap between method and truth, whereas the sceptic denies that we have the epistemic means to bridge the gap. I will argue that neither anti-realist strategy yields an acceptable account of scientific knowledge of an objective world. The internalist strategy loses sight of reality, while the sceptical strategy fails to provide a sustainable account of the relation between evidence and theory.

Internal realism

Internal realism is characterized by an epistemic conception of truth. On such a conception, truth is identified with satisfaction of criteria of epistemic appraisal. According to Hilary Putnam, for example,

> 'Truth', in an internalist view, is some sort of (idealized) rational acceptability – some sort of ideal coherence of our beliefs with each other and with our experiences *as those experiences are themselves represented in our belief system* ... (1981, pp. 49–50)

Similarly, for Brian Ellis, 'truth is what is right epistemically to believe' (1990, p. 10). It 'is what it is ultimately right for anyone to believe, given [our natural] system of [epistemic] values' (1990, p. 11). Thus, according to internal realists, for a claim or theory about the world to be true is for it to be ideally justified or for it to maximize epistemic value.

For the internalist, there is an analytic or conceptual relation between method and truth. Truth consists in appropriate satisfaction of epistemic norms. Accordingly, no problem arises for the internalist of an epistemic gap between method and truth. A theory which is ideally justified, or which maximizes epistemic value, just is a true theory. Nor does any problem arise relating use of the scientific method to advance on truth. If use of scientific method leads to theories which increasingly satisfy the rules of method, it follows immediately that science advances on truth. Given that truth consists in satisfaction of the rules of method, an increase in the level of satisfaction of such rules constitutes advance on truth.

The trouble with internal realism is that it is an inherently idealist doctrine. The epistemic conception of truth entails the mind-dependence of the states of affairs that make our claims about the world true. This may be shown by means of the T-scheme:

(1) 'P' is true iff P.

Given the internal realist identification of truth with epistemic justification, to be true just is to be epistemically justified. Hence,

(2) 'P' is true iff 'P' is epistemically justified.

From (1) and (2), it follows that:

(3) 'P' is epistemically justified iff P.

This means that the state of affairs that P obtains just in case the claim 'P' is epistemically justified. Thus, what (3) says, in effect, is that the existence of a truth-making state of affairs depends on it being epistemically justified to believe that the state of affairs obtains.

This reveals the idealism at the heart of internal realism. If truth is epistemic justification, the states of affairs that make claims true necessarily fail to be objective, mind-independent states of affairs. To revert to an earlier example, suppose it is true that electrons have negative charge. For the internalist, this means that electrons have negative charge just in case we are epistemically justified in believing that electrons have negative charge. But this has the consequence that electrons only have negative charge if we are justified in believing that they do. Thus, for the internalist, the way the world is is not something that is independent of what we think. Rather, the way the world is depends on our being justified in thinking that it is a certain way. Despite promising to span the epistemic gap, internalism therefore fails to provide an account of how scientific knowledge of an objective world is possible.

Scientific scepticism

While the internalist adopts an optimistic view of the relation between method and truth, the view of the scientific sceptic is a decidedly pessimistic one. Both van Fraassen and Laudan maintain that scientists may have good grounds for the acceptance of theories, but deny that rational credence extends to the truth of the transempirical content of theories. Thus, both authors defend a selective scepticism which denies theoretical knowledge while granting credence to observation.

For van Fraassen, the purpose of the scientific enterprise is not to discover truth, but to construct theories that are empirically adequate:

> Science aims to give us theories which are empirically adequate; and acceptance of a theory involves as belief only that it is empirically adequate. (1980, p. 12)

A theory is empirically adequate, according to van Fraassen, 'exactly if what it says about the observable things and events in the world is true – exactly if it "saves the phenomena"' (1980, p. 12). Van Fraassen does not deny that theories make truth-valued assertions about unobservable items. What he denies is that empirical evidence may provide support for the truth of such claims about unobservables.

For his part, Laudan holds that scientific theories may be epistemically warranted, but denies that such warrant extends to their truth. In Laudan's view, 'knowledge of a theory's truth is radically transcendent' (1996, p. 195). Laudan contrasts the transcendent property of truth, which he takes to be 'closed to epistemic access', with other properties which he considers to be 'immanent', such as well-testedness, predictive novelty, and problem-solving effectiveness (1996, p. 78). The principal basis for his rejection of a warranted presumption of theoretical truth lies in his historical critique of the convergent realist claim that there is a correlation between the success of theories and their reference and approximate truth which is best explained by realist means. For Laudan, the fact that there is no way to bridge the epistemic gap between method and theoretical truth is simply a hard fact of the history of science.[7]

The trouble with the sceptical denial of an epistemic connection between method and truth resides in the attempt to combine metaphysical realism with the possibility of a limited epistemic warrant for theories. The scientific sceptic allows that there may be epistemic grounds that warrant acceptance of a theory, but denies that such warrant extends to the truth of the transempirical content of theory. But the sceptic does not deny that scientific theories are capable of being true. Indeed, neither van Fraassen nor Laudan provide grounds for denying that there are facts about the world which make our theoretical claims about the world true or false.

But it is not possible both to allow that theories are made true or false by the way the world is, and to deny that evidential support extends to the theoretical content of theories. If empirical facts about the world are capable of providing evidential support for theories, then such evidential support cannot be restricted to the non-theoretical content of theories. The reason has to do with the nature of the relationship between the empirical facts which provide support and the theories for which such facts provide support.

Scientific theories make claims about both observable and unobservable states of affairs. Among the claims which theories make about observable states of affairs are predictions of observable phenomena that are made on the basis of hypotheses about unobservable portions of reality. In the case of evidence based on the confirmation of such predictions, the predicted phenomena are events that, according to the theory, are brought about by unobservable causal processes. Because such observable events are supposed to be produced by unobservable causal processes, the evidence derived from such observable events has direct relevance to the theoretical hypotheses upon which the predictions of such phenomena are based. Indeed, given that hypotheses about unobservable processes may be the sole basis for prediction of the observable phenomena, the non-empirical content of the theory is directly implicated in the evidential relation between observed fact and warranted theory.

In view of the failure of scientific scepticism to adequately account for the relation between evidence and theory, and the idealism inherent in internal realism, I conclude that neither position provides an acceptable account of scientific knowledge of an objective reality. I will now present the outlines of the scientific realist theory of the relation between method and truth that I propose.

A realist theory of method

The realist theory of method that I propose consists of three key components:

> Epistemic naturalism: normative epistemological questions about rational justification are empirical questions about the best means of conducting inquiry into the objective natural world.

> Methodological instrumentalism: the rules of scientific method are 'cognitive tools' or 'instruments of inquiry', which serve as means for the realization of epistemic ends.

> Abductive realism: the best explanation of the cognitive and pragmatic success of scientific theory and practice is that the

rules of scientific method are genuinely truth-conducive tools, which serve as reliable means for obtaining truth.

These three elements of a realist theory of scientific method form part of a generally naturalistic, non-anthropocentric picture of the world, and of our epistemic relationship to it. We find ourselves embedded in a natural world which we did not create, and over whose fundamental character and structure we have no control. In order to survive, we must form beliefs about the world, and causally interact with it by means of action that is guided by such beliefs. Given the independence of reality from thought, the beliefs that we form about the world do not necessarily correspond to the way that the world in fact is. In such a world, we do not know in advance of inquiry how to proceed to insure survival. Nor can we know by a priori means how best to pursue inquiry into the nature of reality. Thus, the question of how to learn about the world is a question about the contingent nature of our epistemic capacities and the relation of such capacities to the world. Such a question is an empirical question that can only be answered on the basis of empirical investigation into the nature of inquiry.

More specifically, on the instrumentalist conception of method that I favour, the rules of method are to be understood as means for the achievement of epistemic ends. In this I follow Larry Laudan, who has argued that the rules of method may be construed as empirical claims about means to ends. In particular, they may be expressed as hypothetical imperatives of the form, 'If one wishes to achieve aim A, then one should employ method M'. For example, Popper's rule against *ad hoc* hypotheses may be formulated as the hypothetical imperative, 'If one seeks well-tested theories, then one should avoid *ad hoc* hypotheses'.[8]

The instrumentalist construal of method reveals how the normative rules of method may be subject to empirical evaluation. If method M is proposed as a means of achieving epistemic aim A, then it is an empirical question whether use of M reliably conduces to realization of aim A. For example, as Laudan argues, one may consult the history of science for evidence that use of a given method in the past has led reliably to the achievement of given epistemic aims. Thus, the instrumentalist conception of method illustrates the ability of epistemic naturalism to account for the normative force of rules of scientific method. Because the rules of method may be treated as empirically evaluable means to epistemic ends, the epistemic warrant of such rules may be grounded in empirical facts about the nature of inquiry. As such, the normativity of the rules of method derives from empirical facts of procedural efficacy and reliability.

The problem remains, however, of the relation between method and truth. One cannot directly observe that use of the rules of scientific method leads to true scientific theories. The truth of the non-observational content of theories transcends empirical verification, hence cannot be established by direct observational means. It is at this point that appeal is to be made to the scientific realist argument that realism is the best explanation of the success of science. But where the success argument is usually employed to argue for the approximate truth of theories, I extend the argument to the truth-conduciveness of rules of method. I will now sketch the position of abductive realism, which seeks to bridge the gap between method and truth.

Abductive realism

On the scientific realist picture that I propose, the relation between method and truth is not an analytic, conceptual relation, as the internal realist suggests, but a synthetic, empirical relation. It is a contingent relation between epistemic means and ends, which may be known in the a posteriori manner suggested by epistemic naturalism. But the attempt to combine a naturalistic account of epistemic warrant with the realist view of truth as the aim of science must face the problem that no empirical evidence may show directly or conclusively that use of a methodological rule yields theoretical truth. In the absence of direct or conclusive evidence, why should use of a rule of method be taken to conduce to truth?[9]

This is where abductive realism enters the picture. In the absence of direct or conclusive evidence linking method to truth, the grounds for such a link may be at best abductive ones. More specifically, the realist claim that application of rules of method leads to progress toward truth rests on an inference to the best explanation of scientific success. What best explains why scientific theories satisfy the rules of method is that they are close to truth.

Suppose, for example, that there is some theory which satisfies a broad range of rules of method to an extraordinarily high degree. The theory is supported by all available evidence. It successfully predicts a great many previously unknown and surprising novel facts. It unifies previously disparate domains. And it does all of this in a manner which maximizes simplicity and coherence. Clearly, any theory which so impressively satisfies the rules of scientific method is a highly successful theory indeed.

How is such success to be explained? Where a theory impressively satisfies a broad range of methodological rules, the best explanation of such success is that the theory provides an approximately true description of the way the world is. In light of such success, we may infer not only that the entities postulated by the theory exist in roughly the form stated by the theory, but that the underlying causal mechanisms and processes described by the theory really do bring about observable events in the general manner specified by the theory.

It is important to emphasize that the level of descriptive accuracy to which such an inference is committed is that of approximate truth only. While the precise natures of the postulated entities, mechanisms, and processes may fail to be known either in detail or in their entirety, it may nevertheless be the case that such entities, processes, and mechanisms really do exist, in a form which is close to that described by the theory. Given such approximate accuracy, it must also be emphasized that the theoretical description of the postulated entities, mechanisms, and processes remains open to possible revision in the light of further inquiry.[10]

But my point is not simply that the best explanation of the success of a theory, as measured by satisfaction of methodological rules, is the approximate truth of the theory. The crucial point relates to the truth-conduciveness of methods rather than to the approximate truth of theory. Given the critical role played by the rules of method in the process of theory selection, the implications of the success of science for the approximate truth of theory apply with equal force to the rules of method themselves.

In particular, the rules of method are employed by scientists to eliminate theories that are unlikely to be true in favour of theories that are likely candidates for truth. Since the best explanation of satisfaction of rules of method is the approximate truth of theory, and since the rules of method play a critical role in arriving at such approximately true theories, it follows that use of the rules of method is responsible for arriving at theories that are approximately true. Given this, the best explanation of the role played by the rules of method is that the rules are employed in a rigorous selection process which eliminates false theories in favour of theories that are closer to the truth. That is, the rules of method are rules that 'screen for truth' – or, in other words, the rules of method reliably conduce to truth.

Such an abductive realist account of the truth-conduciveness of method has implications as well for the realist view of scientific progress as convergence on truth. Suppose there is a sequence of scientific theories which displays an increasingly high level of satisfaction of the rules of method. According to abductive realism, increased satisfaction of the rules of method is to be attributed to convergence on truth. Where a sequence of theories displays an increasingly high level of satisfaction of the rules of method, the best explanation is that the sequence of theories is advancing progressively closer to the truth.

Thus, on the view I propose, what best explains satisfaction of the rules of method is that the rules are truth-conducive, and what best explains increased satisfaction of such rules is convergence on truth. It is in this sense that I wish to claim that what is needed to bridge the gap between method and truth is an abductive argument to the best explanation of the success of science.[11]

Realism or sheer luck?

It is, of course, a legitimate question why truth and approximate truth should play the role which I ascribe to them as the best explanation of satisfaction of rules of method. To demonstrate that satisfaction of the rules of method is best explained by the truth-conduciveness of such rules would require an exhaustive elimination of alternative explanations. I cannot undertake that task here. But it is instructive to consider a stark anti-realist alternative that is contrary to the abductive realist thesis that the rules of method are truth-conducive rules whose use promotes the discovery of truth about the world. By eliminating this alternative a large and particularly salient class of anti-realist alternatives may also be eliminated.

To generate such a contrary to abductive realism, let us consider the following scenario. Consider, as we did before, a scientific theory which impressively satisfies a great variety of methodological rules. The theory is descriptively accurate and well confirmed by all observational tests. It predicts surprising novel facts in an accurate and reliable manner. It unifies phenomena from domains previously thought to contain disparate and unrelated phenomena. On top of all this, the theory is also maximally simple and coherent.

This time, however, let us also suppose that despite impressively satisfying all the methodological rules the theory is in fact totally and utterly false at the transempirical level. None of the unobservable entities, mechanisms, or processes postulated by the theory exist. Moreover, the theory erroneously imposes unity on unrelated domains which in fact have nothing in common. In short, let us suppose that the theory satisfies all empirical and formal methodological constraints to a very high degree, yet at the level of the descriptive accuracy of its claims about the underlying nature of reality it is simply false.

If such a situation were to obtain, it would be sheer luck that the theory has any success at all. This may be seen most clearly in the case of predictive success, and, in particular, in the case of accurate and reliable prediction of previously unknown and otherwise entirely unexpected phenomena. Either predictive success of this kind is the result of sheer luck, or else there is some benevolent force whose action makes the theory's predictions turn out to be true despite the fact that the transempirical claims of the theory are completely false.

There are, I suppose, possible worlds in which lucky guesses are routinely rewarded with predictive success. But we do not live in such a world. Occasional guesses may succeed. But if a scientific theory reliably produces accurate predictions of novel facts, the best explanation of such predictive success is not that we live in a world that rewards luck. The best explanation is that the theory is at least an approximately correct description of the unobservable entities whose behaviour underlies the observed phenomena predicted by the theory. For this reason, we may conclude that satisfaction of methodological rules provides a reliable indication of advance on truth. The rules of method are a guide to the truth. They are a guide to the truth, not in the sense that truth consists in satisfaction of the rules of method, but in the sense that a theory that satisfies such rules has a good chance of being at least approximately true. If a theory which satisfies the rules of method did not have a good chance of being at least approximately true, the satisfaction of the rules of method would be completely inexplicable.

Notes

1   Acknowledgements: work on this chapter commenced while I held visiting positions at the Center for Philosophy of Science at the University of Pittsburgh and the Centre for Philosophy and Ethics of Science at the University of Hanover. I am grateful to the Directors and staff of both centres for support and hospitality. Earlier versions of this chapter were presented as papers at the University of Roskilde, Humboldt University of Berlin, University of Melbourne, and La Trobe University, as well as at the AAP conference at Melbourne and the international conference on 'The Problem of Realism' in Genoa. I am particularly grateful for comments to John Bigelow, Steve Clarke, Michael Devitt, Brian Ellis, Allen Hazen, Michael Heidelberger, Paul Hoyningen-Huene, Bruce Langtry, Graeme Marshall, Michele Marsonet, Alan Musgrave, Nick Rescher, Jack Smart, and Barry Taylor.
2   The methodological pluralist approach outlined here owes much to the work of Feyerabend (1975), Kuhn (1970), and Laudan (1984). As is well known, methodological pluralism gives rises to the spectre of epistemological relativism. I have sought to dispel this spectre elsewhere. In this chapter, I am concerned with the relation between method and truth, rather than the nature of rational justification or the variation

of the rules of method. For further development of the methodological pluralist approach specifically as it relates to the issue of epistemological relativism, see Sankey (1997 and 2000).

3   The point that the T-scheme is not a definition of truth is made by Tarski: 'neither the expression (T) itself (which is not a sentence, but only a schema of a sentence), nor any particular instance of the form (T) can be regarded as a definition of truth' (1943/1994, p. 110). It might perhaps be thought that a deflationary conception of truth such as Horwich's minimalism does treat the T-scheme as a definition of truth. But Horwich himself notes that deflationism 'does not provide an explicit definition, but relies on a schema to characterize the notion of truth' (1994, p. xv).

4   My insistence that the realist conception of truth requires that claims about the world be made true by mind-independent states of affairs raises the question of the status of claims about mental states and artifacts. Since minds do not exist independently of minds, and artifacts are the product of intentional human action, claims about minds or artifacts would seem incapable of being true in the realist sense. Yet presumably the realist should allow that, at least in principle, such claims might be true. To adequately address this concern would require an analysis of the concept of independence of the mental on the basis of which it may be said that claims about mental states or artifacts are made true by states of affairs that obtain independently of the mental in the appropriate sense. No such analysis can be provided here. But, fortunately, the issue may be set aside for present purposes. The kinds of claims about the world that are of principal concern here are the observational and theoretical claims of the natural sciences. I take it to be highly plausible indeed to say that such claims are made true (or made false) by the way things stand in the world independently of what we humans think about the matter.

5   Cf. Ellis (1990, p. 187) and Putnam (1978, p. 127).

6   When I speak here of the 'internal realist strategy', I mean to restrict attention specifically to the internal realist epistemic conception of truth which defines truth in terms of method or rational justification. As Brian Ellis has pointed out to me, internal realism properly understood is a substantive metaphysical position which is not restricted to an epistemic conception of truth. In particular, the internal realist position is a neo-Kantian position which denies epistemic access to a realm of noumenal objects, and treats objects, reference, and reality as relative to conceptual scheme. Suffice to say that it is the relation between method and truth, rather than any more substantive metaphysical views, that are of relevance for present purposes.

7   See Laudan (1981), reprinted as the final chapter of Laudan (1984).

8   For the suggestion that methodological rules may be construed as hypothetical imperatives, see Laudan (1996, pp. 132–4). Laudan's hypothetical imperative analysis of methodological rules has been the subject of searching criticism (for example, Doppelt 1990, Siegel 1990). In my view, the most challenging objection relates to the source of the epistemic normativity of methodological rules within the hypothetical imperative analysis. The rules of method may only derive normative force by way of the goals toward which such rules are directed. But the source of the epistemic value of such goals remains mysterious. The hypothetical imperative analysis, considered strictly as such, provides no basis on which to evaluate the epistemic merits of any particular epistemic goal. This problem is resolved within the framework adumbrated here by treating truth as the ultimate goal of scientific inquiry from which the value of lower order epistemic goals is derived. For further discussion of this approach, see Sankey (2000).

9   The objection that there may be no direct or conclusive evidence to the effect that use of
    a rule of method conduces to theoretical truth is due to Laudan (1984, p. 53; 1996, p.
    261, n. 19). While I agree that there may be no direct evidence that use of a rule of
    method yields theoretical truth, I hold that there may be indirect evidence that use of
    such rules leads to truth at the transempirical level. For development and defence of this
    idea, see Sankey (2000).
10 In holding there to be a reasonably clear sense of 'approximate truth' which relates to
    the general ontological claims of theory, and does not require explication by means of a
    technical concept of verisimilitude or closeness to truth, I follow Ernan McMullin's
    discussion (1984, pp. 35–6, and 1987, pp. 59–60).
11 The strategy described here as 'abductive realism' is not without precedent in the
    epistemology of science. Broadly understood as inference to the best explanation of the
    success of science applied at the level of method, the strategy is employed by such
    authors as Boyd (1984, pp. 58–9), Kornblith (1993, pp. 41–2), and Rescher (1977, p.
    81ff). For discussion of related uses of inference to best the explanation, see Day and
    Kincaid (1994, pp. 271–3).

References

Boyd, Richard (1984), 'The Current Status of Scientific Realism', in Leplin (1984), pp. 41–
    82.
Day, Timothy and Kincaid, Harold (1994), 'Putting Inference to the Best Explanation in its
    Place', Synthese, 98, pp. 271–95.
Doppelt, Gerald (1990), 'The Naturalist Conception of Methodological Standards in
    Science: A Critique', Philosophy of Science, 57, pp. 1–19.
Ellis, Brian (1990), Truth and Objectivity, Oxford, Blackwell.
Feyerabend, Paul (1975), Against Method, London, New Left Books.
Horwich, Paul (ed.) (1994), Theories of Truth, Aldershot, Dartmouth.
Kornblith, Hilary (1993), Inductive Inference and its Natural Ground, Cambridge, MA,
    MIT Press.
Kuhn, Thomas S. (1970), The Structure of Scientific Revolutions, Chicago, University of
    Chicago Press.
Laudan, Larry (1981), 'A Confutation of Convergent Realism', Philosophy of Science, 48,
    pp. 19–49.
Laudan, Larry (1984), Science and Values, Berkeley, University of California Press.
Laudan, Larry (1996), Beyond Positivism and Relativism, Boulder, Westview Press.
Leplin, Jarrett (ed.) (1984), Scientific Realism, Berkeley, University of California Press.
McMullin, Ernan (1984), 'A Case for Scientific Realism', in Leplin (1984), pp. 8–40.
McMullin, Ernan (1987), 'Explanatory Success and the Truth of Theory', Scientific Inquiry
    in Philosophical Perspective, Nicholas Rescher (ed.), University Press of America, pp.
    51–73.
Putnam, Hilary (1978), Meaning and the Moral Sciences, London, Routledge and Kegan
    Paul.
Putnam, Hilary (1981), Reason, Truth and History, Cambridge, Cambridge University
    Press.
Rescher, Nicholas (1977), Methodological Pragmatism, Oxford, Blackwell.

Sankey, Howard (1997), *Rationality, Relativism and Incommensurability*, Aldershot, Ashgate Publishing.

Sankey, Howard (2000), 'Methodological Pluralism, Normative Naturalism and the Realist Aim of Science', in Robert Nola and Howard Sankey (eds), *After Popper, Kuhn and Feyerabend: Issues in Recent Theories of Scientific Method*, Dordrecht, Kluwer, pp. 211–29.

Siegel, Harvey (1990), 'Laudan's Normative Naturalism', *Studies in History and Philosophy of Science*, 21, pp. 295–313.

Tarski, Alfred (1943/1994), 'The Semantic Conception of Truth', in Horwich (1994), pp. 107–41.

van Fraassen, Bas (1980), *The Scientific Image*, Oxford, Oxford University Press.

Chapter Five

The Real Distinction between Persons and their Bodies

Christopher Hughes

Are my body and I two different things, or one and the same thing? It might once have been thought that how one answered this question would depend on whether one was a materialist or a dualist. Recently, though, there has been something of a consensus that persons may not be identified with their bodies, even if persons are 'nothing over and above' their bodies – even if all the parts of persons are parts of their bodies.[1] Two sorts of arguments have often been offered in favour of the distinctness of persons from their bodies – the argument from temporal discernibility, and the argument from modal discernibility. The argument from temporal discernibility could be put this way:

I and my body do not exist at all the same times.

_____

I and my body are not the same thing.

The argument from modal discernibility could be put this way:

I and my body do not exist in all the same possible circumstances.

_____

I and my body are not the same thing.

In what follows, I shall argue that neither of these arguments refutes the thesis that people are identical to their bodies.

I

The argument from temporal discernibility is clearly valid; but why suppose that it is sound? Why suppose that I and my body do not exist at all the same times? One line of thought would be:

When I die, I will cease to exist. But the same does not go for my body: it will continue existing (as a dead body, or corpse) until it decomposes, or is cremated, or the like. Since my body will outlast me, I and it do not exist at all the same times.

Although this has persuaded many philosophers, a few have doubted or denied the claim that bodies (ordinarily) outlast the persons they embody. Thus Bernard Williams writes:

> Aristotelian enthusiasts will point out that, leaving aside immortality, when Jones (or his body) dies, Jones ceases to exist ('he is no more'), while his body does not. There may be something here: but it surely cannot be pressed too hard. For, taken strictly, it should lead to the conclusion that 'living person' is a pleonasm, and 'dead person' a contradiction; nor should it be possible to see a person dead, since when I see what is usually called that, I see something that exists. And if it is said that when I see a dead person I see the dead but existent body which was the body of a sometime person, this rewriting seems merely designed to preserve the thesis from the simpler alternative that in seeing a dead body I see a dead person because that is what it is. (Williams 1973, p. 74)

It is not immediately clear why someone who believes that Jones actually goes out of existence at death, before his body does, is committed to the claim that 'dead person' is a contradiction in terms. Perhaps it is a contingent fact about Jones that he went out of existence when he died, and never existed as a dead person. Or perhaps Jones could not have existed as a dead person but there are – or at any rate could have been – other persons that could have existed as dead persons. Jones will not become a ghost after he dies; perhaps he could not possibly become a ghost after he dies (if, say, he is essentially corporeal). Still, it might be that there could have been ghosts, and that if there had been such things, they would have been persons who started out as (embodied) live persons, and later became (disembodied) dead persons.

In the passage just cited, Williams implies that 'dead person' should not be assimilated to noun phrases such as 'negative natural number'. This seems right. There is not – and could not be – anything that could truly be described as a negative natural number. But, it seems – at least as long as the existential quantification is not understood as restricted to currently existing individuals – there are individuals that can truly be described as dead persons. Suppose you and a friend are talking about Bob. I come into the conversation late, and ask you who Bob is. You might reply that Bob is the dead man who used to live next door. (Actually, you would be more likely to say that Bob is the man who died recently, and used to live next door; but I take it that if Bob is a man who died recently – and he has not come back to life – then Bob is a dead man.) Again, although it is not and could not be true that there are more negative natural numbers than there are, I have heard it said that there are more dead persons than there are live persons. Still, this does not tell against the view that persons actually, or even necessarily, go out of existence when they die. Someone who thinks that existing while dead is an impossibility can agree that there could be, and indeed are dead persons, although there are not and could not be negative natural numbers. (She just needs to take the quantification as ranging over past as well as present individuals.)

All the same, people do say things like 'I saw a dead person by the side of the road'. How can they truly say things like that, unless some person exists and is dead at the same time? Proponents of the view that persons cease to exist when they die typically answer that the dead person by the side of the road is not a person, any more than a shoe tree is a tree, or a papier mâché lion is a lion; the dead person by the side of the road is the dead body of a person who is no more.

Although Williams suggests that this move is unmotivated, it does not seem so to me. (As long as they have been well and truly shredded) shredded documents are not documents, puréed carrots are not carrots, and fossilized leaves are not leaves. In all these cases, what we call an F-ish K is not a K that is F, but rather what is left of a K after it goes out of existence (sometimes as a result of being F-ed, sometimes as a result of something that happened earlier). So it does not seem *ad hoc* to suppose that 'dead person' means 'what is left of a person after her death' (namely, her body). Or rather, in the sentence 'I saw a dead person by the side of the road' it means that. In a sentence like 'For all those years, she has (unwittingly) been writing letters to a dead person', it means 'person who has ceased to live (in the right sort of way, and has not come back to life)'.[2]

Fred Feldman finds this unpersuasive.[3] After all, he says, in biology classes schoolchildren dissect frogs. Now you can write to someone who does not exist at the time, but you cannot dissect someone (or something) that does not exist at that time. The frogs the schoolchildren dissect accordingly exist – in an ex-animate state – when they are dissected. So why should not the man (or the person) you saw by the road exist in an ex-animate state when you saw him?

Also, suppose that after a person dies, his family buries him. How could one bury someone or something that did not then exist? If I die and someone buries me, what they bury will be me in an ex-animate state (presuming that I cannot be alive unless my body is alive).

These sorts of considerations do, I think (at least initially) increase the plausibility of the view that persons and animals do not (typically) go out of existence when they die. But, *pace* Feldman, I do not think they 'decisively establish' that it is contrary to common sense to suppose that persons or animals (typically) cease to exist when they die (Feldman 1992, p. 95).

Papier mâché lions are not (real) lions. Suppose, though, that you are looking at a bunch of different papier mâché animals in a shop window. You might tell the sales assistant you wanted to buy the lion and the giraffe. You do not want to buy a (real) lion, but there's no impropriety in what you said, since the context makes it clear it's a papier mâché lion, rather than a real one, that you want.

Someone who thinks that persons and animals go out of existence when they die might say that in the frogs example something similar is going on. The things the schoolchildren are dissecting are not really frogs; they are dead frogs (and pickled frogs). But in a normal context, it is perfectly proper for the teacher to say that the children in her class are dissecting frogs, since it is common knowledge that what children dissect are dead frogs.

As for the burial case, suppose that Jones ate fish. This might have happened in a number of ways. Perhaps Jones caught a very small fish, and ate it alive, in one gulp. Perhaps Jones caught a good-sized fish, cooked it, boned it, and cut it up in small pieces before he started eating. In the latter case, I take it that even someone who thinks that a fish goes on existing after death as an ex-animate piscine body will agree that at the time Jones starts to eat, the fish is no more. (If the body of the fish still exists even after its parts are sundered, when does it go out of existence?)

If Smith says 'Jones caught a fish and ate it', he does not thereby commit himself on whether Jones cut up his fish before he started eating. Suppose Brown said to Smith: look, if Jones caught a fish and ate it, he could not have cut the fish up first, since you cannot eat a thing at a time unless it exists at that time. Smith would quite properly accuse Jones of pedantry.

The moral is that one could naturally and properly (relative to non-pedantic standards of propriety) say that someone caught a fish and ate it, even if what he ate was not *sensu stricto* a fish, but something left behind after the fish (and, in this case, after the fish's body) went out of existence. *Pari ratione*, it might be said, one could naturally and properly say that the murderer killed White and buried him in the back garden, even if what the murderer buried was not strictly speaking White, but something left behind after White went out of existence.[4]

Feldman or Williams might protest here that, even if one can explain away ordinary ways of talking that apparently imply that I shall exist as an ex-animate body, it is a demerit of the view that bodies outlast persons that it requires such explaining away.

True. But I think that various things we ordinarily say, and ordinarily do not say, suggest that we think that ex-animate bodies are not the persons that used to have them.

Suppose for example that Smith has just had very bad news about his prospects from the doctor. He might well say to his friend:

> I've just talked with the doctor, and it looks as though I'm not
> going to be here much longer.

By 'here' Smith probably does not mean (say) 'London'. He means that he is not going to be in the world much longer. (Or perhaps, if Smith believes in an afterlife, he means that he is not going to be in this world much longer.) When exactly does Smith think he is going to stop being in the (or in this) world? When he dies. We would be very surprised if, after saying 'I'm not going to be here much longer', Jones added:

> unless of course, I ensure that after my death my body is very
> well preserved – in which case I'll be here for donkey's years.

This certainly suggests that we ordinarily think that after people die they are not there any more, even if their (dead) bodies are.

Also, suppose that Brown's friend White had been planning to go to Nepal last Thursday, but went into a coma on Wednesday (a coma the doctors are sure he will come out of). Suppose that another friend of Brown knew about White's travel plans, but not about the coma. He calls Brown, and says: 'I need to talk to White; is he still in England?' Brown might naturally answer:

> Yes, but he's in a coma.

Now fill in the story as before, but suppose that on Wednesday White died. When Brown's friend says 'I need to talk to White; is he still in England?', I think Brown would be very unlikely to answer:

> Yes, but he's dead,

even if Brown knows that White's dead body still exists, and is in England. If, however, White exists for as long as his dead body does, it is hard to see why 'Yes, but he's dead' should be any less natural or appropriate an answer (in the second scenario) than 'Yes, but he's in a coma' (in the first).

Inasmuch as our propensities to say (and not say) certain things seem not to assort with the view that persons end up as ex-animate bodies, this provides at least some motivation for explaining away the things we say that suggest they do. Granted, Feldmanians may run the same argument in reverse. At the end of the day, I am not sure to what extent I can argue for the view that animals and persons (or at least, persons with no 'asomatic' principle of life such as an immaterial soul) cease to exist at death. It just seems very plausible to me, though not completely beyond doubt, that the dead body that will still exist after my death is not me – is not this person or this human animal. It is not, in its lumpish inanimateness, the right kind of thing to be me.[5]

If the reader does not resonate to this intuition, she may come to it by thinking about when human persons come into existence. If I am an entirely corporeal being, it seems at least initially natural to suppose (as Feldman does) that I came into existence at conception.[6] (After all, one might say, how can I be conceived at a time when I do not yet exist?) But considerations involving (the prehistory of) embryonic development have led some people to think otherwise.[7]

Before the second week, the cells 'descended from' the fertilized ovum adhere to each other only loosely, and have as it were separate lives, metabolizing and dividing independently of each other. The cells are functionally speaking interchangeable, and most will develop into the placenta or other structures that support the embryo, rather than the embryo itself. Separating the clump of cells into two clumps will result in identical twins, although if the two clumps are put back together (within the period before cell-specialization), we will end up with only one human being. Rearranging the cells in the clump will have no effect on future development.

At about sixteen days, though, the cells in the clump undergo specialization and begin to grow and function in a coordinated way. The clump of cells will begin to exhibit bilateral symmetry around the ancestor of a spinal cord (the so-called 'primitive streak'), and splitting the clump of cells into two smaller clumps would result in death, rather than in two embryos.

According to many embryologists, it is at this point that I come into existence. As A. McLaren puts it:

One can trace back directly from the newborn baby to the foetus, and back further to the origin of the individual embryo at the primitive streak stage in the embryonic plate at sixteen or seventeen days. If one tries to trace back further than that there is no longer a coherent entity. Instead, there is a larger collection of cells, some of which are going to take part in the subsequent development of the embryo and some of which aren't. To me the point at which I began was at the primitive streak stage. (McLaren 1986, p. 22)[8]

According to McLaren, before the two-week point, we have not got me, just a 'collection of cells', some but not all of which will take part in the development of the embryo at a later time. Why do the pre-embryonic cells – or at least the pre-embryonic cells that will subsequently take part in the development of the embryo – not constitute me even before embryogenesis? After all, I, as I am now, 'grew from' those cells, just as I, as I am now, grew from the cells that constituted me when I was six years old.

A natural thought here is that the pre-embryonic cells from which I grew do not constitute me before embryogenesis because (as we have seen) they are not sufficiently coordinated in their activities to be regarded as taking part in my life.[9] In this respect, the pre-embryonic cells are very unlike the cells that constituted by body when I was six, and more like the sperm and egg pair from which I came, which obviously never took part in my life. (Even though I 'grew from' the sperm and egg pair, few would be tempted to suppose that the sperm cell and egg cell jointly constituted me prior to my conception.)

Suppose that the pre-embryonic cells from which I grew did not constitute me before embryogenesis because they do not (then) take part in my life. Then, it would seem, neither do the cells of my dead body constitute me after my death. After I die, various things may happen to the cells in my body. Some or all may die along with my body, in which case they will not take part in anyone's life. Some may be transplanted into someone else's body (as parts of a transplanted organ) in which case they will take part in someone else's life. There are other possibilities. But whatever happens, they will not be taking part in my life after my death.

Now if the cells in my body after I die will not constitute me then, then I will not exist as an ex-animate body; and if nothing other than cells of my body will ever constitute me (no soul, no ethereal entelechy, no computer hardware) then I will not exist after my death, just as I did not exist before embryogenesis.

If I will not end my days as a dead body, may we conclude that the argument from temporal discernibility is sound? Feldman appears to think so,[10] but I have my doubts.

What happens to my body when I die? Feldman (together with many other philosophers) consider it obvious that it goes on existing, as a no-longer-living body.[11] Aquinas would not agree.[12] As he sees it, when a person dies, the dead body he leaves behind is not identical – *numero* or even *specie* – with the body he had while he lived.[13] There is the man's (living) body, and there is the man's dead body (which is not really a body at all), and there is the matter that first constituted the living body, and then constituted the dead body. But there is nothing which is (as opposed to constitutes) a body, and first was living, and then was dead (as opposed to: first constituted something living, and then constituted something dead).

This view may seem less strange when we see that it is a counterpart of a commonly held view about persons. Suppose that a living person becomes a dead person (in the sense of 'becomes' in which Lot's wife became a pillar of salt, and in the 'dead person by the side of the road' sense of person). Then, non-Feldmanians will typically say, there is the (live) person, and there is the dead person (which is not really a person at all), and there is the body that first constituted the live person, and then constituted the dead person, but there is nothing which is (as opposed to constitutes) a person, and was first alive, and then dead.

Someone might object here that dead bodies surely are bodies; as often as not, when we talk about bodies without qualifying them as dead or alive, we have dead ones in mind. But, a defender of Aquinas might say, our willingness to say things like 'There's a body in the road', when there is a corpse in the road, no more shows that dead bodies are bodies than our willingness to say 'The children are dissecting frogs today' shows that dead frogs are frogs. Still, could I not truly say, pointing to a dead body 'this body, which is now dead, was once alive?'

Call the (essentially living) body Aquinas thinks I have my body$_L$ (where 'L' stands for living). Call the (accidentally, and probably only temporarily, living) body Kripke and Feldman think I have my body$_{DOA}$ (where 'DOA' stands for 'dead or alive'). Aquinas thinks that there are no bodies$_{DOA}$; by his lights, the belief that there are bodies$_{DOA}$, as well as or instead of bodies$_D$ and bodies$_D$ (that is, corpses) is like the belief that there is 'wineorvinegar' – something that is accidentally and temporarily wine, and accidentally and temporarily vinegar – as well as or instead of wine that turns into (and is replaced by) vinegar.

Whatever the attractions of this view, it is not one that everyone who identifies persons with their bodies must hold. Someone who thinks that persons are their bodies can hold that both bodies$_L$ and bodies$_{DOA}$ are perfectly respectable entities, that 'my body' is ambiguous between 'my body$_L$' and 'my body$_{DOA}$', and that I am my body (that is, my body$_L$).

It may be worth underscoring that this construal of person-body materialism does not trivialize the doctrine. A philosopher who believed in bodies$_L$ might have various reasons for thinking they are distinct from the persons that have them. She might, for example, think that two persons could exchange bodies$_L$, so that persons and their bodies$_L$ were discernible with respect to modal properties. Or she might be a dualist and a believer in the afterlife, in which case she would deny that persons and their bodies$_L$ exist at all the same times.

Someone might object to the construal of person-body materialism under discussion in either of two ways. He might deny that there are such things as bodies$_L$. Alternatively, he might allow that there are or might be such things, but deny that the word 'body' can be used to pick them out.

Why should bodies$_L$ not exist? Eric Olson reports that neither of the German words for 'body' (*Korper* and *Leib*) can be applied to corpses (*Leichen*).[14] (According to the *OED*, *Leib* is in fact derived from the German for 'life'.) If Olson is right, neither *Korper* nor *Leib* means 'body$_{DOA}$', and it is at least initially plausible that those terms mean 'body$_L$'. So why are bodies$_L$ not perfectly respectable entities, unambiguously picked out by certain German sortals?

Of course, it might be that the English term 'body' never means 'body$_L$', just as the German term 'Leib' never means 'body$_{DOA}$'. In that case – even on the assumption that there are such things as bodies$_L$, and that persons are identical to them – it will not be true that persons are their bodies.

In fact, though, I am doubtful that 'body' can never mean 'body$_L$'. A dead frog (I think) is not really a frog at all. In the same way, I can see someone saying, a dead body is not really a body at all – in the relevant (biological) sense of 'body';[15] it is too unlike the (living) things that are uncontroversially bodies (in the biological sense) to count as a body (in that sense) (see note 5). If someone did say that, I would not respond: you are simply using the term 'body' incorrectly. In this connection, it is interesting to note that under the very first heading for the word, the *OED* defines 'body' as 'the whole material organism viewed as an organic entity'.[16]

At this point, someone might object:

> Suppose that 'body' can mean 'body$_L$'. It still won't be true that persons are their bodies, because bodies$_L$ sometimes outlast the persons that had them. Suppose someone's cerebral cortex is completely destroyed, so that her higher brain functions are lost forever. Suppose also that the rest of her brain and body are more or less intact. In such a case, the person goes out of existence, but her body$_L$ does not.

The objection raises large issues, to which I cannot do justice here.[17] But I would answer it by denying that, in the case at issue, the person goes out of existence when the cerebral cortex is destroyed.

Whether or not I can go on existing after my life has ended, I cannot start existing after my life has already begun. But, I want to say, my life began long before my higher mental functions came into operation.[18] (My life was under way at the primitive streak stage – roughly a fortnight after conception – and my cerebrum did not come into existence until at least the sixth week; it did not become operational as an organ of thought until much later than that.) So it certainly looks as though I existed as a not yet thinking being. If so, why could I not exist as a no longer thinking being, after my cerebral cortex had been destroyed?[19] If my becoming a *res cogitans* is not the beginning of my life, why should my ceasing to be a *res cogitans* be the end of my life? Perhaps being a person entails being a thinking being, so that I was not always a person, and (maybe) will not always be a person. But the claim that persons are their bodies$_L$ is perfectly consistent with the claim that bodies$_L$ (and persons) were not always persons, and will not always be persons.

II

If all this is right, then argument from temporal discernibility is not as effective against person-body materialism as is sometimes supposed: it does not refute (every construal of) the claim that I am my body. What about the argument from modal discernibility?

The argument from modal discernibility differs from the argument from temporal discernibility in that its validity is considerably more controversial. A respectable minority of philosophers holds that it does not follow from the fact that *a* might have existed without *b*, that *a* and *b* are distinct.[20] But suppose we grant the validity of the argument from modal discernibility. Should we accept its soundness?

I and my body do not exist in all the same possible circumstances if either there is a possible circumstance in which my body exists, but I do not, or a possible circumstance in which I exist, but my body does not. For reasons already adduced, it seems doubtful that we should find a clear case of a possible circumstance in which my body exists, but I do not. But many philosophers have found it clear that there are possible circumstances in which I exist but my body does not.

For example, certain philosophers have thought that whether or not I actually will undergo (instantaneous) disembodiment at death, I could undergo such disembodiment.[21] That is, they have thought it possible, and compossible with the actual history of the world up to the time of my death, that at the first moment of my *post mortem* existence, I shall have no bodily parts – and, *a fortiori*, none of the bodily parts I had just before my death. It seems clear, though, that my body's having, at the first moment of its *post mortem* existence, none of the bodily parts it has just before my death, is not compossible with the actual history of the world up to the time of my death. (How could one and the same body have one set of bodily parts up to my death, and a completely different set of bodily parts at the next moment of its existence?) If this is right, then there are possible circumstances in which I exist, but my body does not.

But how do we know that I could lack, at the first moment of my *post mortem* existence, all the bodily parts I had just before I died? At this point, Cartesians typically appeal to the conceivability of my doing so. The difficulty is that – as Kripke has emphasized – the fact that one can apparently conceive of something does not seem to guarantee that it is genuinely possible (much less compossible with the history of the world up to now). If you believe that Hesperus is a different planet from Phosphorus, you may see no difficulty in conceiving that something is midway between Hesperus and Phosphorus.[22] If you believe that heat is caloric, and do not believe in molecules, you may see no difficulty in conceiving of a body being hot, even though no molecules are in motion. If you do not believe in atoms, you may see no difficulty in conceiving of water existing, even though atoms do not. Even if you believe that everything is made of atoms, you can still probably apparently conceive of (our current physical theory being massively in error) water existing, even though atoms do not.

None of these examples show that the conceivable is sometimes impossible. There is at least some inclination to say that if something is genuinely impossible, it is not genuinely conceivable either, and when we think we are conceiving it, we are actually conceiving something else (which is genuinely possible).[23] But this does not help the Cartesian. If apparent conceivability and real conceivability come to the same thing, and the conceivability or otherwise of a state of affairs (by a person) can be ascertained a priori (by that person), then conceivability is not a guarantee of possibility, and the Cartesian has to explain how we know it is not just conceivable, but also genuinely possible, that I could survive the instantaneous loss of all my material parts. If conceivability is a guarantee of possibility, then the conceivability or otherwise of a state of affairs (by a person) cannot be ascertained a priori (by that person), and the Cartesian needs to explain how we know it is not just apparently conceivable, but also genuinely conceivable, that I could survive the instantaneous loss of all my material parts.

Here the Cartesian might object:

> As Kripke has argued, we have overriding reasons to suppose it is impossible for there to be something midway between Hesperus and Phosphorus, or for there to be heat but no molecular motion, or for there to be water but no atoms. It is accordingly unreasonable for us to believe that any of those things are possible. But we have no overriding reasons to deny that I could survive the instantaneous loss of all my material parts. Given that we can apparently conceive of that state of affairs, it is reasonable for us to believe that it is possible.

But it is not always reasonable to believe that an apparently conceivable state of affairs is possible, even in cases where we have no overriding evidence of its impossibility. Suppose an ancient astronomer has no idea whether Hesperus and Phosphorus are the same celestial body or different celestial bodies. Then, it seems, he can apparently conceive of there being a celestial body midway between Hesperus and Phosphorus. (He might say: it is conceivable that there is such a body.) Should the astronomer conclude that it is possible for there to be a celestial body midway between Hesperus and Phosphorus? Well, suppose he can also apparently conceive of Hesperus being Phosphorus. (He might say: it is conceivable that Hesperus and Phosphorus are one and the same.) Suppose also that, as a Kripkean *ante litteram*, our astronomer believes that if Hesperus = Phosphorus, then it is impossible that Hesperus ≠ Phosphorus, and impossible that something be midway between Hesperus and Phosphorus. Then he should not believe that it is (genuinely, as opposed to epistemically) possible that there be something midway between Hesperus and Phosphorus. He should reason as follows:

> Suppose that although I can apparently conceive that *P*, I can also apparently conceive that *Q*, and I can see that if *Q* is true, then it is not genuinely possible that *P*. Then the fact that I can apparently conceive that *P* does not by itself make it reasonable for me to believe it is genuinely possible that *P*. In the absence of independent grounds for thinking that *Q* is false – that is, grounds that do not already presuppose that *P* is possible – it would be unreasonable for me to believe that it is genuinely possible that *P*. In the case at hand, I can apparently conceive that there is something midway between Hesperus and Phosphorus. But I can also apparently conceive that Hesperus = Phosphorus; and I can see that if Hesperus = Phosphorus, then it is not possible that Hesperus ≠ Phosphorus, and not possible that there is something midway between Hesperus and Phosphorus. Moreover, I have no independent grounds to believe it is false that Hesperus = Phosphorus. So it would be unreasonable for me to believe it is genuinely possible that there is something midway between Hesperus and Phosphorus.

In much the same way, a materialist might say:

> On the face of it, we can at least apparently conceive of my having material parts at every moment before my death, and existing without material parts at the first moment of my *post mortem* existence. That is because there is no a priori detectable entailment between 'I exist' and 'I have material parts': my concept of a self (or of my self) does not exclude (partial or complete) immateriality. But, on the face of it, we can also at least apparently conceive that I am a being all of whose parts are material. Again, there is no a priori detectable incompatibility between 'I exist' and 'I have (only) material parts': my concept of a self (of my self) does not exclude (complete) materiality, any more than it excludes (partial or complete) immateriality. Now I can apparently conceive that the history of the world is just as it actually is up to my death, and I have no material parts at the first moment of my *post mortem* existence. Let us grant that this provides me with a defeasible reason to think that it is genuinely possible that the

history of the world is just as it actually is right up to my death, and I have no material parts at the first moment of my *post mortem* existence. On the other hand, I can also apparently conceive that the history of the world is just as it actually is right up to my death, and I have only material parts at every time before my death. If I have only material parts throughout my (*ante mortem*) life, then for me to come to have a purely immaterial constitution after my death would be for me to have one set of parts right up to my death, and a completely different set thereafter. But nothing can have one set of parts right up to time *t*, and a completely different set of parts at the first moment of its existence after *t*. So, let R = the history of the world up to my death is just as it actually is, and I will have no material parts at the first moment of my *post mortem* existence. Let S = I have only material parts right up to my death. Although I can apparently conceive that R, I can also apparently conceive that S, where (i) I can see that S is true, then R is not genuinely possible, and (ii) I have no R-independent reasons for thinking that S is false. (Appealing to the possibility of my surviving the instantaneous loss of all my material parts – together with the premiss that purely material things cannot survive the instantaneous loss of all their material parts – obviously violates the independence requirement, and is no better than arguing from the possibility of there being something midway between Hesperus and Phosphorus – and the necessity of identity, and the necessary irreflexivity of the '___ is midway between ___ and ___ relation' to the falsity of Hesperus = Phosphorus.) So, by the principle formulated earlier, it is not reasonable for me to believe that R is genuinely possible. In other words, it is not reasonable for me to believe that I could survive instantaneous dematerialization at death.

In nuce, the difficulty the Cartesian faces is this: given the assumption that purely material things could not instantaneously lose all their parts, that I could survive the instantaneous loss of all my material parts and that all my parts are material cannot both be compossible with the history of the world up to now; how can the Cartesian show that the former state of affairs has a better claim to be regarded as possible than the latter?

The Cartesian might say that although we can (at least apparently) conceive of my surviving the instantaneous loss of all my material parts, we cannot (even apparently) conceive of my being made of nothing but matter, any more than we can (even apparently) conceive of a material object's being made of nothing but holes.

This is a hard saying. I can (at least apparently) conceive of a scientist starting out with nothing but material parts, and putting (just) those parts together to get something that looks (not just at the macro level, but also at the micro level) exactly like me. I can also (at least apparently) conceive that the very thing he makes in the laboratory has a mind, is the subject of experiences, and so on. So, it seems, I can at least apparently conceive of a (human) person with a purely material constitution. Or, for those whose tastes do not run to science fiction, I can at least apparently conceive that one material being (a sperm cell) fertilizes another (an egg cell) and sets off a sequence of biological events which, although it does not involve any immaterial beings, nevertheless results in the existence of a human person. So again, I can at least apparently conceive of a human person with a purely material constitution. And if I can do that, I can at least apparently conceive of my having a purely material constitution, since I can at least apparently conceive that the sperm and egg story is true of me.

Alternatively, the Cartesian might say that although we can apparently conceive that I have only material parts, there are independent grounds for thinking that I do not. The difficulty is in seeing what those grounds might be. If the Cartesian knew, he would not need a modal argument to establish that I am not a (purely) material being.

III

We have been considering an argument to the effect that we are not justified in supposing that I might undergo instantaneous dematerialization at death. A crucial premiss of that argument is that a thing cannot have one set of parts right up to t, and a completely different set of parts at its first moment of existence thereafter.[24] In fact, though, some philosophers who hold that persons and their bodies are distinct will reject this principle. For such philosophers, if a machine copied all the 'information' in my brain onto an (initially) 'blank brain' in a body cloned from my body, and destroyed my (old) body and brain, I would survive, even though my old body and brain did not.[25] Right up to the time at which the information (or a sufficient portion thereof) left my brain, I would have one body and set of bodily parts; at the next instant of my existence, I would have a different body, and completely different set of bodily parts. If the machine effected the transfer of information instantaneously, I could go from having one body and set of material parts just before midnight, to having a different body, and completely different set of material parts at midnight. (And if the machine instantaneously effected the transfer of information to something immaterial, I could go from having a body and material parts just before midnight, to having no body and no material parts at midnight).

If I really could be 'resited' in another body via information transfer, then – subject to the general caveats about modal discernibility arguments mentioned earlier – I am not my body (since my body could not be 'resited' in another body). The difficulty is that, on the face of it, copying all the information in my brain onto an (initially) blank brain in another body would turn something (or someone) else into a (psychological) duplicate of me, in much the way that (photo)copying all the information on one sheet of A4 paper onto another (initially) blank sheet of A4 paper would turn the initially blank sheet of paper into an (imperfect) physical duplicate of the original sheet of paper. True, the sort of machine that Shoemaker *et alii* have in mind not only copies the information in my brain onto an initially blank one in a new body; it also destroys the brain the information was copied from (and the body the brain was a part of). So that sort of machine both turns something else into a (psychological) duplicate of me, and destroys my brain and body. But why should we suppose this would result in my continuing to exist? If having the information in my brain copied onto a new brain is, as it were, neutral with respect to my continuing to exist, and destroying my body and brain is (at least in ordinary circumstances) prejudicial to my continued existence, why should doing both of those things at once afford me a way of continuing to exist (albeit in a new body)?[26]

IV

So there does not seem to be any good reason to suppose that I could have my current body instantaneously replaced by a new one. But even if I could not have my body instantaneously replaced by another one, it does not follow that I could not have it gradually replaced by a new one. After all, a liver presumably cannot survive the instantaneous replacement of all the cells constituting it at a given time, but it could survive their gradual replacement; a ship cannot survive the instantaneous replacement of all its planks, but it could – it seems – survive the gradual replacement of all those planks.

Cannot we in fact imagine scenarios in which my body is gradually replaced by a different body, or by some kind of inorganic replacement for my body? After all, I can survive the replacement of (small-ish) parts of my body by inorganic substitutes – for example, the replacement of one of my teeth by an artificial tooth, or the replacement of my hip joint by an artificial one. As long as the parts replaced were sufficiently small, and the replacement were gradual enough, could I not survive the replacement of all the current parts of my body (including all the parts of my brain) by inorganic substitutes? It looks as though I can imagine such a thing happening to me. And if it did happen to me, would I not end up with a new (a different) body?

Someone who wants to identify me with my body might respond in either of two ways. She might concede that I could survive the (gradual) replacement of all my parts by inorganic parts, but insist that the same is true of my body. The idea would be that my body is only accidentally organic (even if it is, presumably, essentially originally organic).[27]

It seems doubtful that my body could survive the replacement of all its organic parts by inorganic ones. Suppose that a dentist gradually replaced more and more of one of my teeth with amalgam, until finally, there was no organic material left at all, just amalgam. I take it that this would not result in my original (originally entirely organic) tooth becoming entirely inorganic; rather it would result in my original (organic) tooth being replaced by an artificial one. Similarly, though I do not know how to argue the point, it seems to me the gradual replacement of all my organic body parts by inorganic ones would result in my original body being replaced by an artificial one.

Alternatively, someone who wants to identify me with my body might concede that if all my organic parts were gradually replaced, so would my body, but insist that the same holds for me. The idea would be that at some point in the sequence of gradual replacements, I would go out of existence, and some new thing, spatio-temporally and causally continuous with me would come into existence.

Why is it hard to believe that the gradual replacement of all my organic parts by inorganic ones would necessarily result in my ceasing to exist? I think the answer is something like this: the concept of a mind seems unlike the concept of, say, a tooth or a heart. Our concept of a tooth or of a heart is a concept of something organic, so that inorganic teeth are false teeth (or teeth in a different sense (as when we talk about the teeth of a gear)), and an inorganic heart is an artificial heart, not a real one. The same goes for our concept of a brain. By contrast – even if all minds are in fact organic – our concept of a mind is not a concept of something organic: we do not feel that God or angels would have 'false minds' or artificial minds, or minds in a different sense. (To put this another way, although our concept of a brain is – like our concept of a heart – the concept of an organ of a certain sort, our concept of a mind is not the concept of an organ of a certain sort: theists do not think of God as having organs, even though they think of Him as having a mind.)

Just as our concept of a mind is not a concept of an organ of thought, it is not the concept of something causally related (in the right sort of way) to something organic. (Again, if it were, having a mind could not be univocally predicated of God and us.)

Thus our concept of a mind leaves room for the possibility that a mind could start out as something with an organic constitution, causally related in certain ways to something else with an organic constitution (namely, the rest of the body), and end up as something which neither had an organic constitution, nor was causally related in those ways to anything with an organic constitution.

Furthermore, we have a strong inclination to think that the survival of my mind is a sufficient condition for my survival. If, however, my mind could become something inorganic, not causally related (in the right sort of way) to anything organic, and if my mind survives only if I do, then I could survive the replacement of all my organic parts by inorganic parts, and the replacement of my current body by an inorganic substitute, if the replacement of all my organic parts by inorganic ones would amount to the replacement of my current body by an inorganic substitute.

These considerations constitute an intuitive case for the claim that my surviving the replacement of all my organic parts by inorganic ones is (at least conceptually) possible. (Or, to put it more cautiously, they constitute an intuitive case for the claim that there is no obvious conceptual incoherence in my surviving the replacement of all my organic parts by inorganic ones.) On the other hand, the claim that I could survive the replacement of all my organic parts by inorganic ones seems in tension with other intuitions.

Nothing can be an animal without being an organism, and nothing can be an organism without being made of organic parts. So if I could come to be made of completely inorganic parts then either (1) I am not an animal (although I as it were 'share a body' with an animal), or (2) I am an animal only contingently. Although Locke is arguably committed to (1), even those who do not want to reject it out of hand generally concede that it is (initially, at least) counter-intuitive.[28] Now (2) is not as obviously counter-intuitive as (1), but the proponent of (2) has to face some hard questions. It is a plausible (Aristotelian) principle that for each individual, there is a (non-trivial) kind to which it essentially belongs – a kind that provides the answer to the question, 'what (kind of thing) is it?'[29] One might have thought, along with the scholastics, that the kind I essentially belong to is *homo sapiens*. But this cannot be right, if I am only contingently an animal (given that nothing could be a member of the species *homo sapiens* without being a member of the animal kingdom). So what kind of thing am I (essentially)? According to the Cartesian tradition, a thinking thing. But, as we have seen, this does not assort with some widely shared intuitions about when I began to exist. Moreover, as Olson has argued,[30] if I am essentially a thinking thing, then it is hard to see how we can avoid the (unpalatable) conclusion that I and the animal sharing my body are numerically distinct. (Some animal existed in my mother's womb, before it or I had a mind. That animal still exists, and now has the same body I have. So if I am essentially minded, and did not exist until I had a mind, then there is an animal now 'in my body' that existed before I did, and thus cannot be identical to me.) In the face of these difficulties, it might be suggested that I am neither essentially an animal, nor essentially a thinking thing: perhaps I am essentially originally an animal, and essentially either-an-animal-or-a-thinking-thing. Before I acquire a mind my persistence is tied to my going on being an animal of a certain kind; after I acquire a mind, I could persist either by going on being the same animal (even if I lost my mind), or by going on having the same mind (even if I ceased to be an animal).[31] Maybe something like this can be made to work; but it is not obvious that it can, and in any case the idea that I have as it were an irreducibly disjunctive essence is suspect.

V

I have been suggesting that even if my mind could survive the replacement of all my organic parts by inorganic ones, it is not as clear as one might have thought that I could survive such replacement. Thus, even if it is conceded that my mind, but not my body, could survive the replacement of all my organic parts by inorganic ones, it does not obviously follow that I am not my body. But should we concede that it is genuinely possible for my mind to survive the replacement of all my organic parts by inorganic ones? Might not the fact that there is no manifest incoherence in supposing that my mind survives such replacement be a reflection of our ignorance?

Perhaps. But I think that whether or not the scenario we have been considering is genuinely possible, someone who thinks that I am my body will have to address the problem of what would happen to me if my mind went on, but my body – or at any rate, most of my body – did not.[32] Suppose my brain were removed from my skull, and put in a new body-except-for-brain (say, one cloned from my own). And suppose that my brain continued to function in its new environment (that is, continued to generate thoughts, experiences, and the like). Suppose further that after my brain had been removed, the rest of my (original) body was destroyed. What would happen to me? And what would happen to my body?

I doubt that someone who identifies me with my body can refuse to answer this question on the grounds that we are not dealing with a genuine possibility. The case under discussion resembles cases we know are possible (involving transplants of other organs) for us to have any confidence that it is impossible.[33]

If she grants that the case at issue is possible, the advocate of what we may call the 'corporealist' account of (human) persons has at least two options: she may say either that in such cases, my mind outlasts me, or that in such cases, my body survives the loss of all its non-cerebral parts.

We have already considered the drawbacks of the first option. *Pace* Descartes, most of us do not think of ourselves as essentially minded (since we think of ourselves as having acquired a mind after we came into existence); but we do have a strong inclination to suppose that the continued existence of our mind is sufficient (if not necessary) for our continued existence. Still, as we have seen, there may be reasons to resist that inclination.

If the survival of my mind is independent of the survival of the rest of my body, and sufficient for my survival, then it looks as though I could survive, not just as a brain transplanted into a new (all-but-the-cerebral-part-of) a body, but also as a brain in a vat. Inasmuch as the idea that I am essentially an animal has a certain appeal, and inasmuch as it is at least doubtful that an envatted brain is an animal, this constitutes grounds for doubt that the survival of my mind is sufficient for my survival.[34]

Alternatively, the corporealist may hold that when my (living) brain and the rest of my body are separated, the things that were non-cerebral parts of my body cease to be parts of my body, and my body is reduced to something all of whose parts are cerebral parts. On this way of thinking about brain-transplant cases, they are extreme cases of amputation. If someone has a toe removed, she presumably does not lose her old body, or acquire a new (a different) body; she goes on having her old body, although that body has fewer parts than it used to have. Similarly, on the view under consideration, when someone's non-cerebral part of the body is removed, she does not lose her old body, or gain a new one; her (persisting) body loses all its non-cerebral parts. If the brain is subsequently put into a new body-except-for-brain, then she presumably gets a complete new set of non-cerebral bodily parts, but not a new body.[35]

I am not sure what to make of this last suggestion. Suppose that someone could make 'from scratch' a living, thinking, (human) brain in a vat. I would certainly be reluctant to describe the (purpose-built) thing in the vat as having (or being) a human body. But if it would not have (or be) a human body, then presumably neither would something that had been separated from the non-cerebral bodily parts it had been originally joined to, and subsequently envatted. And if the separated and envatted thing would not have (or be) a human body, can we really say that it would still have (or be) its old body – the body that was human before envatment? Could being human be an accidental property of a human body?

On the other hand, we do ordinarily suppose that our bodies can survive the loss of some of their parts. We do not think of the person who has lost a finger as having lost her old body. Even if a person loses a substantial portion of her body – say, both arms and both legs, we do not think of her as having a numerically different body from the one she had before the accident. (Philosophers who think that persons are not their bodies do not support their view by appeal to cases in which people lose all their limbs.) Once it is granted that a human body could survive the loss of a substantial portion of its non-cerebral parts, it is not clear that there is a principled reason to insist that it could not survive the loss of all its non-cerebral parts.[36]

VI

It is sometimes thought that identifying persons with their bodies is no more defensible than identifying gold rings with the gold they are made of, inasmuch as the same kind of (temporal or modal) discernibility arguments that establish the distinctness of a gold ring from the gold it is made of also establish the distinctness of persons from their bodies. I have tried to show that this is not so. Although the identification of persons with their bodies gives rise to certain difficulties, it is not clear that they are insuperable, or for that matter graver than the difficulties encountered by those who distinguish persons from their bodies.[37]

Notes

1   See, for example, Wiggins (1983, p. 162).
2   I say 'ceased to live in the right sort of way' because, if there are non-human persons, some of them might cease to live without dying (if, say, they undergo fission).

3 See Feldman (1992, pp. 93–5). Although the frogs example and the burial example are Feldman's, I have not presented the arguments in quite the form Feldman does. But the gist of the arguments is 'Feldmanian'.

4 This strategy for blocking the 'burial case' argument was suggested to me by an example of Feldman's ('The fish you eat today, last night slept in Chesapeake Bay') which he uses to argue in favour of the idea that (most) animals (sooner or later) exist as dead bodies (1992, p. 95).

5 One reason we might be tempted to think that a (recently) ex-animated human body is a (real) person is that it looks so much like a (real) person. But imagine a machine whose exterior is made of a highly conductive metal with a very high melting point, and whose innards are made of metal with a very low melting point. If, as a result of being subjected to intense heat, the machine's 'innards' are fused, though its exterior is unaffected, I want to say that what is left is not the machine we started with, no matter how much it looks like it (from the outside). Similarly, I want to say, a day-old corpse with disabled and decaying 'micro-innards' is not the person who died a day ago, however much the two might resemble each other from a 'macro' point of view (cf. Olson 1997, p. 152).

   Moreover, at the phenomenological level, I think people find (recently ex-animated) bodies 'creepy' precisely because they think: 'it looks so much like a person – but it isn't one'.

6 If I consist simply of an immaterial soul, or consist of an immaterial soul together with a human body, then there is no obvious reason to suppose that the moment of conception is the beginning of my existence, even if it is the beginning of the existence of my body.

7 See the very helpful discussion of these matters in Olson (1997, pp. 89–93). In what follows I draw on Olson.

8 Cited in Olson (1997, p. 174, n. 10).

9 The connection between being a part or constituent of a living organism and taking part in that organism's life is emphasized by Peter Van Inwagen (1990, p. 91).

10 Feldman says that 'any clear-headed proponent' of person-body materialism (the view that persons are their bodies) 'would undoubtedly say that death usually does not make people cease to exist' (Feldman 1974, p. 667).

11 Thus Kripke writes: '[P]rovided that Descartes is regarded as having ceased to exist upon his death, 'Descartes ≠ B [Descartes' body] can be established without the use of a modal argument; for if so, no doubt [Descartes' body] survived Descartes when [it] was a corpse' (Kripke 1971, pp. 163–4, n. 19).

12 At least, he would not agree if what he means by the Latin *corpus* is what we mean by the English 'body'.

13 See Quodlbetum II, q. 1, a. 1.

14 Olson (1997, p. 151).

15 There is a sense of 'body' according to which things that are not and never have been alive can be bodies (such as celestial bodies). A corpse may perfectly well be a body, in a non-biological sense of 'body'.

16 That 'body' at least used to mean something like 'organism' – and not that long ago – is evident from the (current) meaning of the term 'antibody'. It is not just scholastics or neo-scholastics who think that there is a sense of 'body' in which dead bodies are not bodies: see for example Lowe (1991).

17 I address them at some length in 'Identità Personale e Entità Personale'.

18 As do many embryologists (cf. the passage from McLaren cited earlier).

19 For a vigorous and to my mind persuasive defence of the claim that I existed as a not yet 'minded' fetus, see Olson (1997, ch. 4). For a very crisp expression of the intuitions underlying the claim that I am only temporarily and contingently minded, see Olson (ch. 2, sec. 3).

20 See, for example, Lewis (1983, pp. 47–54), Gupta (1980), and Noonan (1993, pp. 133–47).

21 See, for example, Swinburne (1984).

22 Assume (at least for the sake of argument) that Hesperus could not be midway between itself and itself.

23 Some of Kripke's remarks suggest that he holds this sort of view; see, for instance, Kripke (1971, p. 160).

24 Here 'part' must be understood in a sense more restricted than the one current among mereologists. In the mereologist's sense of 'part', everything is a (maximal) part of itself, so that a thing's survival trivially entails the survival of at least one of its parts.

25 See, for example, Nozick (1981, p. 39); Shoemaker (1990, p. 122).

26 Shoemaker concedes that most of us are initially inclined to think that the sort of machine he describes duplicates a person and destroys the original, rather than resiting the person in a different body; but he offers an argument that he takes to enhance the plausibility of the latter view of what the machine does. I take issue with the argument in 'Identità Personale e Entità Personale'.

27 Someone who takes this line will have to readdress the first argument we considered for the distinctness of persons from their bodies (if my body could go on existing in a non-organic state, it seems not be a bodyʟ in the straightforward sense).

28 Cf. Shoemaker (1984, p. 114).

29 Cf. Shoemaker (1984, p. 114).

30 See Olson (1997, ch. 5).

31 I discuss some of the difficulties in 'Identità Personale and Entità Personale'. Also see Olson (1997, pp. 84–5).

32 Williams, who defends the thesis that persons are bodies against various objections, thinks that it is most vulnerable to objections involving cases where a person's mind/brain and (the rest of) her body come apart: cf. (1973, pp. 76–9).

33 For a defence of the possibility of brain transplants, see Snowdon (1991).

34 Snowdon (1991) argues in just this way.

35 Van Inwagen argues that, in brain-transplant cases, the human animal survives a particularly extreme form of amputation (1990, pp. 172–81.) This is consistent with, but does not obviously entail, that the animal's body survives a particularly extreme form of amputation. More generally, the kind of animalist view of human persons defended by Snowdon, Olson *et alii* (according to which we are essentially animals of a certain kind) is apparently consistent with, but does not obviously entail a corporealist view of human persons (according to which we are identical to our bodies).

36 Of course, when a finger is separated from the rest of my bodily parts, there is no doubt that the rest of my bodily parts, in Nozickian terminology, have a better claim to be the continuers of my (pre-amputation) body than the finger has. When the brain is separated from the rest of my bodily parts, it does not seem obvious that the brain has a better claim to be the continuer of my pre-amputation body than the rest of my bodily parts have. It might be said that since the rest of my bodily parts cease to live upon their

separation from their brain, they have no hope of jointly constituting any bodyL. But what if the rest of my bodily parts are kept artificially 'alive', by being hooked up to a machine that works like an artificial brainstem cum midbrain?

37 Many thanks to Michele Marsonet, Mark Sainsbury, and Wesley Salmon.

References

Feldman, F. (1974), 'Kripke on the Identity Theory', *Journal of Philosophy*, 71, p. 665–76.

Feldman, F. (1992), *Confrontations with the Reaper*, New York, Oxford University Press.

Gupta, Anil (1980), *The Logic of Common Nouns*, New Haven, Yale University Press.

Hick, J. (1990), *Philosophy of Religion*, Englewood Cliffs, NJ, Prentice-Hall, 4th ed.

Hughes, C. (2001), 'Identità personale ed entità personale' in A. Bottani and N. Vassallo (eds), *Identità personale. Un dibattito aperto*, Naples, Loffredo, 2001, pp. 157–97.

Kripke, S. (1971), 'Identity and Necessity', in *Identity and Individuation*, M. Munitz (ed.), New York, New York University Press.

Lewis, D. (1983), 'Counterparts of Persons and Their Bodies', *Philosophical Papers*, vol. 1.

Lowe, E. J. (1991), 'Persons as a Substantial Kind', in *Human Beings*, D. Cockburn (ed.), Cambridge, Cambridge University Press.

McLaren, A. (1986), 'Prelude to Embryogenesis', in *Human Embryo Research: Yes or No?*, London, Tavistock.

Noonan, H. (1993), 'Constitution is Identity', *Mind*, 102, p. 133–47.

Nozick, R. (1981), *Philosophical Explanations*, Cambridge, Belknap Press.

Olson, E. (1997), *The Human Animal: Personal Identity Without Psychology*, Oxford, Oxford University Press.

Shoemaker, S. (1984), 'Personal Identity: A Materialist's Account', in S. Shoemaker and R. Swinburne, *Personal Identity*, Oxford, Blackwell.

Snowdon, P. F. (1991), 'Personal Identity and Brain Transplants', in *Human Beings*, D. Cockburn (ed.), Cambridge: Cambridge University Press, 1991.

Swinburne, R. (1984), 'Personal Identity: The Dualist Theory', in S. Shoemaker and R. Swinburne, *Personal Identity*, Oxford, Blackwell.

Van Inwagen, Peter (1990), *Material Beings*, Ithaca, Cornell University Press.

Wiggins, D. (1980), *Sameness and Substance*, Cambridge, MA, Harvard University Press.

Williams, B. (1973), 'Are Persons Bodies?', in *Problems of the Self*, Cambridge, Cambridge University Press.

Chapter Six

A Realistic Account of Causation

Wesley C. Salmon

My approach to causation begins with David Hume for two reasons. First, Hume's writings on causation constitute the *locus classicus* for modern discussions of this topic. Second, his position contrasts sharply with the view I want to articulate and defend – it is the paradigm of an irrealistic position. At the outset, I must state explicitly the sense of the term 'realism' I am adopting in this chapter. It is not a Platonic realism that asserts the reality of universals, construing causality as relationship among them (for example Fales 1990). It is a common-sense sort of realism – one that accepts the reality of ordinary everyday medium-sized material objects such as billiard balls, baseball bats, children, barns, cigarettes, and windows. It leaves open questions about the reality of such microscopic and submicroscopic entities as viruses and molecules. It also leaves open questions about the status of sense data, which may turn out to be the product of psychological theories rather than a fundamental 'given' of experience. Thus, my point of departure is neither Platonistic nor phenomenalistic.

In addition, I assume that we have a good deal of empirical knowledge about ordinary objects and their behaviour. I presume that we understand the meanings of such concepts as mass, momentum, and energy as applied to these objects. The point of departure is not the mind as a *tabula rasa* or a state of total ignorance. We have knowledge of various empirical generalizations – such as conservation of linear momentum – though I am not granting these generalizations the status of laws of nature. They more closely resemble Hume's constant conjunctions. Humean scepticism regarding basic induction is put aside for present purposes. I readily acknowledge that our experiences of ordinary material objects sometimes fail to be veridical; therefore, we do not have an incorrigible basis in experience. Nevertheless, I aim to show that causation has the same sort of physical status as everyday material objects. I intend to do so without abandoning the principles of empiricism, but also, without being committed to Hume's eighteenth-century psychology.

Hume on causation

In his *Treatise of Human Nature*, Hume (1888 [1739–40]) discussed causation at length. Disappointed by the reception of this work, which, he said in his autobiography, '*fell deadborn from the press*' (1955 [1776], p. 4, Hume's italics), he wrote an anonymous Abstract (1955 [1740]) in which he focused on causation as the central topic. Later, in the *Enquiry Concerning Human Understanding*, he returned to the same subject. Because of the concise and emphatic nature of the Abstract, it serves well as a point of departure. Hume writes:

> Here is a billiard ball lying on the table, and another ball moving toward it with rapidity. They strike; and the ball which was formerly at rest now acquires a motion. This is as perfect an instance of the relation of cause and effect as any which we know either by sensation or reflection. Let us therefore examine it. It is evident that the two balls touched one another before the motion was communicated, and that there was no interval betwixt the shock and the motion. *Contiguity* in time and place is therefore a requisite circumstance to the operation of all causes. It is evident, likewise, that the motion which was the cause is prior to the motion which was the effect. *Priority* in time is, therefore, another requisite circumstance in every cause. But this is not all. Let us try any other balls of the same kind in a like situation, and we shall always find that the impulse of the one produces motion in the other. Here, therefore, is a *third* circumstance, viz., that of a *constant conjunction* betwixt the cause and effect. Every object like the cause produces always some object like the effect. Beyond these three circumstances of contiguity, priority, and constant conjunction I can discover nothing in this cause. (1955 [1740], pp. 186–7)

Hume's account is, of course, more famous for its omissions than its content. He fails to find any logical relations between causes and effects. Given knowledge of a cause, we cannot infer from that alone what the effect will be. Conversely, given knowledge of an effect, we cannot infer from that alone what the cause was. In the famous example of the billiard balls, he cannot infer what will happen when the two balls meet. Perhaps the second ball will remain stationary, and the first ball will reverse its motion and return to its starting place. Hume says that he can imagine this without inconsistency. Is this simply a manifestation of Hume's psychology? Is what he imagines genuinely consistent?

We can, in fact, show that it is. Given Hume's description of the situation prior to the collision, we can add the condition that the motionless ball is firmly fixed to the table by a large screw. Under these circumstances, the second ball will remain motionless after the collision. If there is any doubt about this, we can readily perform an experiment. Obviously, whatever is actual is logically consistent. Hume did not say anything about the ball being screwed to the table, but that does not matter. Take our description of the situation and remove the statement about the screw. It is a simple point of logic that a consistent set of statements cannot be rendered inconsistent merely by removing one or more statements from the set. So Hume is right, as a matter of logic, and not merely as a matter of psychology. The same consideration applies in the opposite direction as well. We can arrange for a hidden lever in the surface of the table to impart a motion to the second ball an imperceptibly short time before the first ball arrives. The same lever, carefully designed, will halt the motion of the first ball just before the collision would have occurred. Thus, from the motion of the second ball, we cannot infer the collision with the first ball. The argument proceeds *mutatus mutandus* as in the case of inferring the effect from the cause. Thus far, Hume's argument is logically impeccable.

Hume not only (rightly) fails to find a logical connection between causes and their effects; he also fails to find any 'necessary connection' between causes and effects, or any 'secret power' by which the cause produces the effect. What he does acknowledge is 'constant conjunction'. As a matter of 'habit' or 'custom', repeated observations of the cause followed by the effect arouse in us an expectation that the same effect will follow, given the next observation of the cause. Causation is not a feature of the physical world known by reason or empirical observation; it is a product of the human imagination. According to Hume, then, causality is no more than what nowadays we call a conditioned response. It is just like Pavlov's dogs. In his famous experiment, Pavlov sounded a bell just before feeding his experimental animals. After this practice had been repeated enough times, the dogs salivated when the bell rang – thus expressing their expectation of food – even when no food was presented.

I am reminded of a famous story – I cannot vouch for its authenticity – about Bertrand Russell and Alfred North Whitehead, the joint authors of *Principia Mathematica*. After Whitehead had moved to Harvard University subsequent to the First World War, Russell was invited to give a lecture there. He chose to speak about ethics, and his talk was filled with references to the scientific psychology of the time, including psychological conditioning and Pavlov's dogs. At the conclusion of the lecture, just as the audience was about to begin applauding, Whitehead, who had introduced Russell, sprang to his feet and declared, 'Verily, I say unto you that The Good is but a certain watering at the mouth'. We can easily adapt

Whitehead's comment: Verily, Hume says unto us that causation is but a certain watering at the mouth.

Reactions to Hume

If Hume's view were correct, it would imply that there can be no such thing as causation in the physical world apart from human expectations. There could not have been causation before people or other intelligent creatures inhabited the world. Causation would cease to exist if all such creatures became extinct. This is a viewpoint we find hard to accept. I am convinced, for example, that the Grand Canyon in northern Arizona was produced – caused – by erosion of the Colorado River over a period of many millions of years, long before any humans or other intelligent forms of life existed on Earth.

Not surprisingly, many philosophers sought to undermine Hume's subjective account. In a twentieth-century classic, *The Cement of the Universe*, J. L. Mackie (1974) reviews some of the most important attempts. I believe that his criticisms of his predecessors are largely valid. He then elaborates his own theory. His goal is to locate causation 'in the objects' rather than 'in the mind' (pp. 1–2). If his programme had been carried out successfully, it would have resulted in a fully objective realistic account of causation. I shall argue, however, that he fails. I take issue with him on three main points.

Logical probability

Mackie gives a detailed reconstruction of Hume's critique of causality, but he claims to have found a fundamental flaw. He agrees that Hume was correct in rejecting the notion that deductive logical connections can be found between causes and effects, but he argues that Hume understandably ignored another possibility, namely, a probabilistic logical relationship between causes and effects (p. 15). This avenue was not open to Hume, because the logical interpretation of probability is essentially a product of such twentieth-century philosophers as John Maynard Keynes (1921) and Rudolf Carnap (1950). It seems to me that this suggestion is fruitless, however, because the introduction of logical probabilities involves the use of arbitrary a priori measures of probability, as Carnap's highly developed system reveals (see Salmon 1967, 1969). If one were to pursue this approach, it would turn out that the existence or non-existence of a causal relationship in a particular situation would hinge on the language chosen to describe it. Such linguistic arbitrariness is just as intolerable as the subjectivity of Hume's account.

Causal regularities

Hume's writings on causality do not make clear whether he takes a cause to be a sufficient condition of its effect, a necessary condition, or both. For present purposes, that does not matter. On any of these construals, Hume is proposing a so-called regularity account, that is, either all *As* are followed by *B*, only *As* are followed by *B*, or all and only *As* are followed by *B*. Each is a universal generalization. I have a deep dissatisfaction with such regularity theories, because it seems natural to ask, in any of these cases, what connection stands between the *As* and *Bs* to sustain the regularity in question.

Although Mackie does not hold a regularity theory, he offers an account in which a rather sophisticated regularity, involving a complex combination of necessary and sufficient conditions, plays a central role. He distinguishes two senses of the term 'cause'. As it is ordinarily used, 'cause' refers to a relationship known as an INUS condition, that is, an Insufficient, Nonredundant (that is, necessary) part of an Unnecessary Sufficient condition (1974, p. 62).

Abstractly, the INUS condition is hard to grasp, but it is easily illustrated. Suppose that a barn burns down.[1] There are various possible causes, for example, a lighted cigarette dropped by a careless smoker, spontaneous combustion in hay stored there, a defect in its electrical wiring, an act of deliberate arson, or being struck by lightning. Suppose, for the sake of illustration, that the dropping of a burning cigarette by a careless smoker is the actual cause. Obviously, this is not a necessary condition, since we have just listed several other possible causes. At the same time, the dropping of the lighted cigarette is not, by itself, sufficient, because other circumstances must obtain. The cigarette must have landed on some combustible material such as dry straw. The straw must be located close enough to other combustible materials for the fire to spread. There must have been no one to put out the fire before it spread. And so on. So, the dropping of the lighted cigarette is a non-redundant part of a set of conditions that are sufficient to produce the fire, but this set is not necessary in view of other possible causes.

Mackie (pp. 63–4) goes on to characterize the 'cause in the philosophical sense', that is, the full cause, as a large disjunction of all of the sufficient causes, in which each sufficient cause is a conjunction of all the parts required to achieve sufficiency. In the case of the barn, one of the disjuncts is the dropping of the cigarette, the presence of dry straw, its proximity to other combustible materials, and so on. Another disjunct would be spontaneous combustion, that is, storage of moist hay, leaving it undisturbed for a long time, bacterial action, failure to put sufficient salt on the hay to combat the bacterial action, and so on. The result is a proposition in disjunctive normal form that Mackie claims to be necessary and sufficient for the effect in question. This necessary and sufficient condition must be taken against the background of a causal field of standing conditions, such as the presence of oxygen and the absence of asteroids colliding with Earth at that particular place.

Mackie makes the reasonable admission that the choice of one factor rather than another among the INUS conditions as the cause in the ordinary sense depends on all sorts of pragmatic or contextual considerations (p. 73). Furthermore, by referring the full cause to a causal field, he makes the same concession regarding the cause in the 'philosophical sense'. This seems to me to be an admission that he has not succeeded in finding the cause 'in the objects'. The field is a contextual feature, and it depends on such factors as our background knowledge and our interests. Moreover, Mackie admits that the statement of necessary and sufficient conditions is an 'elliptical or gappy universal' – one that contains blank spaces for the insertion of as yet unknown sufficient conditions (p. 76). It is not a statement; it is the logical form of a statement that would express a necessary and sufficient condition of the effect in the presence of a causal field. Immediately after acknowledging this point, he gives a detailed defence of such statements, but his defence rests on their practical utility, not on their mind-independent objectivity (pp. 67–75).

Causal priority

Mackie worries that his account in terms of INUS conditions may fall victim to the lack of causal asymmetry he found in Mill's theory. In Mill's theory, as in Hume's, the direction of causality is determined solely by the time order of events. I agree that this is an undesirable feature of any theory of causality. To repair the difficulty, Mackie introduces the notions of causal priority and causal fixity (pp. 178–83). The idea is that in the causal order of things, causes may be fixed (that is, fully determined and unchangeable) when the effects are not fixed, but effects cannot be fixed when their causes are not. Whether the temporal priority always agrees with causal priority is a separate issue. Mackie proposes that if A and B are directly causally related, and if there is a time at which A is fixed but B is not, then A must be causally prior to B, and not conversely. Unfortunately, this definition breaks down completely in certain ordinary situations.

In this connection, I have a small story to tell. One evening, many years ago, Nancy Cartwright, John Earman, and I were together in a bar in Oberlin, Ohio, the site of a fine undergraduate college bearing the same name. The Oberlin Philosophy Department sponsored an annual philosophy conference to which the three of us had been invited.[2] While Earman was engaged playing pool with a local shark, Cartwright and I played pinball. The pinball machine is a device in which the player shoots a small metal ball to the top of a sloping board. The ball descends, striking various obstacles that light up and produce a score on that particular play. During the game I suddenly realized that we had before us a counter-example to Mackie's fixity thesis. When the ball reaches the top and begins to descend we have a fixed event A, and this will inevitably lead to a final result C, that is, the ball coming to rest at a particular location at the bottom of the board. Event C is fixed from the moment the ball begins its descent from the top. However, there are many different paths B_i by which the ball may travel to the bottom; the particular path the ball will follow on a given play is not fixed from the beginning. According to Mackie's fixity thesis, then, the ending condition C is causally prior to the particular path B_i the ball takes to reach the bottom. Since there are, very often, various different ways to arrive at a given end, this objection is quite general. Consequently, Mackie's fixity condition does not adequately establish causal priority.

Obviously, Mackie has not dealt with every theory of causation proposed since Hume's time, but he has surveyed several of the leading candidates (including counter-factual ones: ch. 2).[3] Moreover, he has constructed, as an essential part of his own theory, as sophisticated a regularity account as seems possible. Nevertheless, it fails to provide a fully objective, context-free explication of causation. In my view, the time has come to propose radically different approaches. In what follows, I shall elaborate what seems to me the most promising available avenue.

A new point of departure

Hume's characterization of causation involves a pair of events (or facts) C and E, the cause and effect in question, and some relation R holding between them. Most subsequent approaches have adopted the same pattern; the crucial question then concerns the nature of this causal relation R. I propose that we temporarily stop thinking about separate facts or events, and that we temporarily eliminate the terms 'cause' and 'effect'. Instead, I suggest that we begin with the notion of a process.[4] We are all familiar with many examples of processes. For instance, a moving material object is a process; in fact, so is a particle that is at rest in our particular frame of reference.[5] A moving shadow is a process; so is a red dot that moves on a projector screen because of a laser pointer used by a lecturer. The propagation of a wave is a process. A photon moving in space is a process. To put it roughly, a process is something that has an extended world-line in a space-time diagram. Events, roughly, are represented by dots in space-time diagrams. So processes are relatively extended in time, and perhaps in space as well. Events are relatively localized things such as a sneeze, a bullet hitting a point on a target, or a collision of two automobiles.

A natural question at this point is how large or small processes and events may be. The answer is highly pragmatic – it depends on the nature of our investigations. To a traffic engineer, for instance, a moving automobile may be a single process. To an automotive engineer, who is concerned with the operations of automobile engines, a moving automobile would be a complex system of processes and interactions. To an astronomer, a planet such as Earth moving in an orbit around the sun might be a single process.[6] To a geophysicist, it would be an extremely complex system of processes and interactions.

I shall put my cards on the table right away. I want to show that causal processes are precisely the causal connections Hume tried in vain to discover. In some cases, however, they will not be necessary connections; probabilistic connections are also admissible. In order to make the case, it is first necessary to show how we can distinguish causal processes from pseudo-processes. That is the task to which I now turn.

Processes: causal versus pseudo-

The most important feature of causal processes is that they transmit something effective – for example, energy, information, electric charge, momentum, or causal influence. Pseudo-processes do not. The question is how to characterize the difference. Two answers have been given, one by Hans Reichenbach (1956 [1928]), the other by Phil Dowe (1992).

According to Reichenbach, causal processes have the capacity to transmit marks. Here is a simple example of a type often used by Reichenbach and later by me. In my office hangs a beautiful poster showing La Lanterna, the famous lighthouse of Genoa.[7] From the window of my hotel room in Genoa I had a lovely view of it. When it is illuminated at night, it is a rotating beacon, casting its beams of white light in different directions as it turns. When the beam encounters an opaque object such as the outer wall of the hotel, or distant clouds near the horizon, a moving spot of white light appears on the surface. The light that travels outward from the beacon is clearly a causal process. As Reichenbach pointed out, if you put a red filter in the beam at any place, the beam of light travelling outward from the beacon will become red and continue to be red from that point on, without any other marking interactions. You can, of course, make the spot falling on the opaque surface red at any point – for example, by placing a piece of red cellophane there – but as the spot of light moves on, it will return to its white colour, unless you continue marking it as it moves along from one place to another. The transmission of marks by causal processes is familiar. A radio transmitter sends out a carrier wave that can be marked by modifying its amplitude (AM) or its frequency (FM). The transmission of such marks obviously makes it possible to transmit information. Such processes make it possible (when the equipment is working correctly) for NASA to influence the behaviour of its space vehicles. Physiological processes can be marked by inserting radioactive tracers; this enables physiologists to learn how substances are dispersed within the body of an organism.

Much more recently, Dowe (1992) has proposed a 'conserved quantity' theory of causal transmission, which has certain important advantages (for purposes of philosophical analysis) over the mark method. Some familiar conserved quantities are energy, momentum, and electric charge. The first advantage of the conserved quantity theory is that it defines causal processes in terms of characteristics that they actually possess (namely, one or more conserved quantities) instead of a mere capacity (namely, the ability to transmit marks). Other advantages of the conserved quantity theory will emerge when we discuss causal interactions.

Another fundamental difference between causal processes and pseudo-processes arises from the special theory of relativity. According to that theory, no signal can travel at a speed greater than the speed of light in a vacuum. A message transmitted by visible light or any other type of electromagnetic radiation obviously travels at the speed of light.[8] Material objects, which might be used to convey messages, can be accelerated to speeds approaching that of light, but never equal to or greater than that of light.[9] Pseudo-processes, in contrast, can travel at arbitrarily high velocities.[10]

Recall the example of the lighthouse. Let us imagine surrounding it with a circular solid wall, whose radius is equal to the distance from the lighthouse to my Genoa hotel. The lighthouse casts a spot of light that travels around this wall in the time it takes for the source of light to rotate once – say one revolution per second. Keeping the beacon at the centre of the circle and its rate of revolution constant, imagine that we expand the surrounding wall to a greater size. Since the spot of light moving along the wall must make a full trip around the circumference in the time it takes for the beacon to rotate once (one second), obviously the speed at which the spot moves must increase. As we keep on expanding the circle we will reach a size that requires the spot to travel at a speed greater than that of light.[11]

For a more extreme example, substitute the pulsar in the Crab nebula for the rotating beacon. Pulsars are believed to be rapidly rotating neutron stars that send out beams of radiation quite analogous to the light from the lighthouse. In fact, the Crab pulsar radiates at optical frequencies.[12] Just as the lighthouse projects a moving spot of light on the outer wall of the hotel, so does the Crab pulsar project a spot of light on our planet. The pulses occur at the rate of thirty times each second; this means that the pulsar rotates at that rate. The Crab pulsar is located 6500 light years from Earth. If we draw a circle around the pulsar with a radius equal to its distance from Earth, we can think of the pulsar sending out a beam that creates a moving spot of light around the circumference of the circle. Light requires 13,000 years to travel across the diameter of this circle, but the spot – definitely a pseudo-process – makes its trip all the way around the circumference in one-thirtieth of a second. As the 'spot' from the pulsar crosses the face of Earth, it is travelling at approximately $4 \times 10^{13} \times c$ (the speed of light).

To help secure our intuitive grasp of the distinction between causal processes and pseudo-processes, let us look at a couple of everyday examples. An aeroplane flying on a sunny day casts a shadow on the ground. The airplane is a causal process; the shadow is a pseudo-process. If two such aeroplanes, flying at different altitudes, pass over a given point on the ground at the same time, their shadows intersect, but the shadows pass beyond the point of intersection as if nothing had happened. If, however, the planes had been flying at the same altitude, and had collided with one another, the result would have been much different – the destruction of the planes and the deaths of all people aboard them.[13]

Another familiar example can be found in the cinema.[14] The actions that we see on the screen are all pseudo-processes. The cowboy digging in his spurs as he jumps on his horse does not cause the horse to start running. The causal processes are located in the projector. Rays of white light pass through the patterns on the film, and they cause the scenes to appear on the screen.[15] The sequence of actions is determined by the successive frames on the film and the motion of the film in the projector. Were an overexcited (perhaps inebriated) viewer to draw a gun and shoot the cowboy just as he is mounting the horse, it would have no effect whatever on the subsequent actions of either the horse or the rider.

In order to explain the distinction between causal processes and pseudo-processes, I have made use of the concept of transmission: the transmission of marks and the transmission of conserved quantities. Because transmission is a patently causal concept, we must now turn our efforts to understanding that notion.

Causal transmission

The distinction between causal processes and pseudo-processes is obviously fundamental to the theory of causation I am attempting to develop, whether we are thinking in terms of marks or of conserved quantities. Let us return again to the lighthouse beacon. As the light rays are emitted from the light source, energy is being transmitted from the beacon to a wall on which the light falls. In contrast, the spot of light travelling along the opaque wall has energy, but it does not transmit energy. Here is the basic difference. As the beam of light is emitted from the beacon, energy is transferred from the light source to the light beam, but no further input of energy is involved as the beam travels from the source to the wall. The spot that moves across the wall also has energy, but it continues to exist and to possess energy only because new energy is provided at each point in its path. In other words, the supply of energy in the spot must constantly be replenished, whereas the energy in the beam travels with the beam without replenishment.

To clarify this concept of transmission, I should like to go back about twenty-five centuries to the famous paradoxes of Zeno of Elea, in particular, his arrow paradox. Zeno apparently argued that, at each moment of its flight, the arrow is at rest. Since, at each moment, it occupies an amount of space precisely equal to its size, it has no extra space in which to move. Moreover, because each moment of time is indivisible, it has no time in which to move. Hence, it is always at rest, so it cannot possibly move.[16]

An immediate modern reaction to this argument is to appeal to the infinitesimal calculus, and to the distinction it allows us to make between instantaneous motion and instantaneous rest. Velocity is a relationship between position and time; the instantaneous velocity is given by the derivative, dx/dt, of position with respect to time. Since Zeno had no access to this branch of modern mathematics, he was understandably unable to grasp the distinction between instantaneous motion and instantaneous rest.

Bertrand Russell (1922), however, pointed out that this response to Zeno is actually question-begging. The derivative function is defined by taking the average velocity over a non-zero span of time $\Delta x/\Delta t$. We form a sequence of such ratios, taking smaller and smaller time intervals Δt. The derivative is the limit of these ratios as Δt goes to zero. Thus, to introduce the derivative at all, we must deny the very conclusion Zeno claimed to have established. Zeno argued that there is no such thing as motion. The foregoing answer amounts to saying, well, let us just assume that things can move. Such a resolution of the arrow paradox is patently unsatisfactory. The difficulty is deeper than it seemed at first sight.

In order to find a satisfactory resolution of the paradox, Russell introduced what he called 'the at-at theory of motion'. We describe motion by pairing various points of space with corresponding moments of time. To say that an arrow moves from A to B means that it occupies each intervening point of space at the appropriate moment of time.[17] Notice that this definition of motion does not involve the notion of instantaneous velocity. It does not say that the arrow zipped through these points very rapidly; it says simply that the arrow was at that place without any mention of velocity. If asked how the arrow moved from the bow to the midpoint of its path, the answer is the same, namely, it occupied the intervening points at the appropriate moments. If asked how the arrow gets from one point to the next, one simply rejects the question because, in a continuous path, there is no next point. Between any two points there are infinitely many others.

Making use of Russell's fundamental insight, we can now apply the 'at-at theory' to causal transmission. To say that a mark is transmitted from A (the point at which it is introduced) to some other place B means that the mark is present at every point between A and B without being reintroduced anywhere along the way. It is at the appropriate place at the appropriate time. To say that a conserved quantity, say electric charge, is transmitted between A and B in a given process means that the given electric charge is present at A and at B and at every intervening point in the process without being resupplied at any intervening place. This formulation is neutral regarding the direction of transmission.

In order to explain the concept of transmission, I have referred to such notions as imposing or reimposing a mark and supplying or resupplying a conserved quantity. These are obviously causal concepts, inasmuch as a causal interaction is involved in the imposition of a mark or the transfer of a conserved quantity. We now need to direct our attention to causal interactions.

Causal interactions

At the very beginning, I introduced the concept of a process, before distinguishing between causal processes and pseudo-processes. At that stage of the discussion, process was not a causal notion; in fact, by referring to space-time diagrams, I suggested implicitly that it is a geometric notion. Following this line of thought, we can look at the intersections of processes as a further geometrical notion; they are intersections of world-lines in space-time diagrams. The task we now face is to distinguish between causal interactions and mere spacetime intersections. For this purpose, we need another basic concept, namely, change. This, too, is a non-causal notion, inasmuch as changes may be caused or uncaused. My aim is to characterize causal interactions in terms of process, intersection, and change.

When two processes intersect, changes may occur in the immediate locus of that intersection. Such changes may or may not persist beyond the locus of intersection. For example, I mentioned the intersection of the shadows of two aeroplanes flying at different altitudes. Although both shadows change shape in the intersection, they continue beyond just as if no intersection had occurred. We shall not classify intersections of this sort as causal interactions. In contrast, when Hume's billiard balls collide, both of these processes change in ways that persist beyond the locus of intersection. The momenta of the two balls are altered. One was originally at rest; its linear momentum was zero. The linear momentum of the other ball was not zero; one possible outcome is that all of its linear momentum was transferred to the ball originally at rest, so that the one initially at rest now moves and the one initially in motion is now at rest.[18]

These two examples give us strong hints about distinguishing between causal interactions and mere spatio-temporal intersections. In the causal interaction between the two billiard balls, each of the two incoming processes is altered in a fashion that persists beyond the locus of intersection. We can regard the interaction between the two billiard balls as an imposition of a mark on each of them, namely, a change in its state of motion. The marks are transmitted, because the altered state of motion persists without any external influence. This example is similar in principle to the marking of the white light by a red filter. In this latter case, the filter is one process and the light beam is another. When they intersect, the colour of the light is changed, and the filter becomes warmer by absorbing energy from the light. In contrast, when the shadows of the aeroplanes intersect, there is no modification in either shadow that persists beyond the locus of the intersection. This intersection is not a causal interaction.

Another way to describe these examples is in terms of the transfer of conserved quantities. When the billiard balls collide, there is a transfer of linear momentum from one to the other. When the light passes through the filter, there is a transfer of energy from the light to the filter. If we think of causal processes in terms of their ability to transmit marks, we can say that a causal interaction is an intersection in which each of the incoming processes is marked (that is, modified) and the mark persists beyond the locus of interaction without any further marking interactions. If we think of causal processes in terms of transmission of conserved quantities, we can say that a causal interaction is an intersection of processes in which the outgoing processes possess amounts of some conserved quantity differing from the amounts possessed by the incoming processes. In contrast, the shadows of the aeroplanes do not possess conserved quantities; the quantities they do exhibit, such as shape and size, are not conserved. *A fortiori*, when the shadows intersect, there is no exchange of conserved quantities.

At this point we can see another major advantage of the conserved quantity approach over the mark method. In the intersections considered thus far, we have had two incoming processes and two outgoing processes. Let us call them 'X-type intersections'. There are, of course, other kinds of intersections. For example, there are cases in which one process splits into two, without any other process impinging. When a hen lays an egg, there is initially a single animal; subsequently, there is an animal and an egg.[19] Another example is the fission of a single-celled organism into two daughter cells. In both of these cases, we can take mass as the conserved quantity. The mass of the hen just before she lays the egg is not equal to her mass thereafter or to the mass of the egg. Of course, the mass of the chicken plus the mass of the egg just after the laying is equal to the mass of the chicken just before the laying, but that is simply a restatement of the conservation of mass – a principle that is surely tenable in this non-relativistic situation. Thinking about a space-time diagram in which time runs from the bottom to the top (as is customary), let us call intersections of this sort 'Y-type intersections'. Another Y-type intersection occurs when an atom in an excited state emits a photon, thus making a transition to a lower energy state. At the outset we have one process (the atom in the excited state) that then splits into two (a photon and a less energetic atom). In this case, energy is the obvious conserved quantity.

Atoms also absorb photons, transforming them into more energetic states. In this case we have two processes merging and coming out as one. Again, energy is the relevant conserved quantity. We can call intersections of this sort 'λ-type intersections'. The considerations mirror those in Y-type intersections. When a snake swallows a mouse, we have another example of the lambda-type intersection; it is strongly analogous to the hen laying an egg. Mass is again the conserved quantity.[20]

It seems to me that the conserved-quantity theory handles the Y and λ intersections quite straightforwardly; when two processes either merge into one or emerge out of one, the conserved quantity in question is different in the incoming and outgoing processes. Thus, a conserved quantity is exchanged. I do not see how the mark method can deal adequately with them. For purposes of philosophical analysis, then, Dowe's conserved-quantity theory is the one I now hold, but the mark method is retained as an effective tool for discovering and verifying causal relationships. The X, Y, and λ interactions are the basic types; obviously, more complicated types exist. For example, when a free neutron decays spontaneously, it yields an electron, a proton, and an antineutrino. A case of this sort can be handled straightforwardly in the same manner as a Y interaction. In fact, I see no reason why there may not be six incoming processes and four outgoing ones, just to pull numbers out of thin air.

One might easily get the impression that the intersection of two causal processes always yields a causal interaction. This would be a mistake. When, for example, two light waves travelling in different directions intersect, there is interference characteristic of waves – when crest meets crest we have constructive interference; when crest meets trough we have destructive interference – but the interference pattern does not persist beyond the locus of intersection. Consider four people sitting at a square table, one person on each side, as in the game of bridge. North and south see one another; east and west see one another. The light rays that make this possible travel on perpendicular paths that intersect. Yet, the light waves that make it possible for north to see south pass through those that enable east to see west, without any lasting distortion in either set of rays. However, it is obvious that there can be no causal interaction between two processes unless both of them are causal processes. Causal processes are necessary for causal interactions, but not sufficient.[21]

Foundations of a realistic account

Now that we have the concepts of causal process, causal transmission, and causal interaction at hand, we have the materials with which to build a realistic account of causation. To show how this is accomplished, I shall recapitulate the discussion in an order more or less reverse to that in which it was presented. The preceding order was heuristic; the recapitulation is constructive. We begin with the non-causal concepts of process and intersection. From the standpoint of space-time diagrams, these are essentially geometric concepts. In the course of the development, we use such straightforward empirical concepts as properties and changes. We introduce quantities as properties of objects or processes; some quantities are, as a matter of fact, conserved.

We are now ready to introduce a sequence of causal concepts. The first of these is causal interaction.[22] It is a space-time intersection of processes. Such an intersection qualifies as a causal interaction if and only if the intersecting processes undergo exchanges of conserved quantities at the locus of intersection that persist for some time beyond that locus in the absence of additional intersections. The persistence need not be long-lasting; it may be quite brief because additional causal interactions may occur frequently. Causal interactions of the X type produce changes, causal interactions of the Y type produce splitting of processes, and interactions of the λ type produce mergers of processes. Causal production is to be identified with causal interactions.

The second causal concept to be introduced is causal transmission. A conserved quantity is transmitted between points A and B in a process if and only if it is present in the same amount at all points between A and B without any additional interactions between these points.[23] Similarly, a change (or mark) is transmitted between points A and B if and only if it is present at all points between A and B without any additional interactions between these points.[24]

The third causal concept is causal process (as opposed to pseudo-process). A process is causal if and only if it actually transmits one or more conserved quantities. A conserved quantity is a property of whatever possesses it. Causal propagation is to be identified with transmission via causal processes.

At the beginning of this discussion, I suggested that we temporarily abandon the terms 'cause' and 'effect'. The time has come to consider reintroducing them. Although the reintroduction may seem simple, we shall see that it takes a bit of doing. Given two events, spatio-temporally separated from one another, under what circumstances can we say that they are directly connected as cause and effect?[25] Although it is rather easy to identify some necessary conditions for a direct cause–effect relation – for example, that there exist one or more causal processes connecting them[26] it is much more difficult to find sufficient conditions. It seems to me that there are at least two basic patterns that we naturally take to qualify as direct cause–effect relations. Examples of both have already been mentioned; let us reconsider them.

The first and simplest is Hume's billiard-ball collision. This is just a causal interaction between two processes. The interaction produces changes in both processes, but because Hume was more interested in the change from rest to movement on the part of the second ball, he said that the collision caused the motion of the second ball. In a slightly different context, we might be more interested in the subsequent motion of the first ball. Suppose the game is pool rather than billiards; the balls used in the two games are the same, except for their markings. The pool table, unlike the billiard table, has pockets. The cue-ball is initially put in motion by the player; the aim is to make it strike one of the object-balls in such a way that the object-ball will drop into a pocket. If, however, the cue-ball drops into a pocket, that is called a 'scratch', and the shooter is subject to a penalty. If a scratch occurs, it would be natural to say that it was caused by the collision with the object-ball – that is, the interaction caused a certain kind of motion on the part of the first ball, the one that was initially in motion rather than the one originally at rest. In an X-type interaction, both of the incoming processes are modified, and both of these modifications are effects of the interaction.

The second type is a bit more complex. This cause–effect pattern consists of a causal interaction between two processes, a causal process that emerges from the interaction, and a subsequent interaction of that process with another process. The interactions can be considered events, so this pattern consists of two events connected by a causal process. For example, a child hits a pitched baseball with a bat, the ball changes its direction and flies toward a window, and then the ball strikes the window causing it to shatter. To quote Hume, 'This is as perfect an instance of the [second type of] relation of cause and effect as any which we know either by sensation or reflection'. In common parlance, the kid hitting the ball with the bat causes the breaking of the window (the effect). In this case, the bat and the ball are two separate processes that intersect and interact causally. The ball is a causal process that moves from the bat to the window. The causal process constituted by the flying baseball intersects and interacts causally with the process constituted by the window pane. Less formally, why did the window break? Because the kids were playing baseball in the vacant lot next door, and Kim hit a fly ball that crashed through the window. The travelling baseball is the causal connection between the striking of the ball and the shattering of the window. This causal process transmits mass, momentum, and energy from one event to the other. It is the causal connection Hume sought in vain.

There is a kind of symmetry in these two examples. Both exhibit two basic features of causality, production and propagation. In the case of the billiard balls, we have two processes, some features of which are produced by a causal interaction. These features are propagated beyond the locus of the interaction by the modified processes. The subsequent motion of one ball is indirectly causally related to the subsequent motion of the other. In the baseball example, we have two causal interactions connected by a causal process. Causal interactions produce changes in causal processes; causal processes propagate the results. My fundamental thesis is that every instance of a cause–effect relation involves a more or less complicated pattern of causal processes and interactions. However, realization of such patterns is patently insufficient for an instance of either type of cause–effect relation. Recall the baseball example. Suppose that, just as the ball is about to strike the window, the batter, anticipating what is about to happen, shouts 'Oh, no!' (or words to that effect), so that the sound waves from the exclamation reach the window just when the ball does. We recognize immediately that, although both the sound wave and the ball carry linear momentum, the momentum of the sound wave is insufficient to break the window. A clear constraint here is the conservation of linear momentum. When we think that a conservation condition might be violated, we withhold the judgment that a cause–effect relation obtains. It appears that something essential is being left out of the story. Non-violation of conservation relations is another necessary condition. If there is any doubt about it in this case, empirical experiments can confirm that baseballs break windows, whereas normal sounds do not.[27]

The problem of context

The major obstacle to the creation of a fully objective and realistic theory of cause–effect relations is the fact that the instances we tend to select are highly context dependent. Take Hume's billiard-ball example again. I have already remarked on the difference between the viewpoint of the player of billiards and the player of pool. This suggests that our interests are a function of the rules of the game, and these are human constructions. Moreover, they depend on the proficiency of the player. In either game, the spin on the cue-ball is a matter of serious concern, because it largely determines what will happen to the cue-ball after the collision. This means that angular momentum, in addition to linear momentum, must be taken into account.[28] The tyro, like Hume's player, tends not to look ahead.

Context also figures conspicuously in the baseball example. As the ball travels from the bat to the window, it undergoes a great number of causal interactions with the molecules along its path through the air. Since the people involved are interested in the general form of the interaction between the ball and the window, not with the precise trajectory of the ball, they will ignore these interactions. If, however, they were thinking about the curve thrown by the pitcher, it would be essential to take interactions with the air into account.

For another example, recall the case of the barn that burned down. Here we have only a more complicated structure involving processes and interactions. A lighted cigarette falls from the fingers of a passing tramp onto some dry straw on the floor. In the interaction between these two processes, energy is exchanged, raising the temperature of the straw to its ignition temperature. The burning piece of straw releases energy that ignites neighbouring pieces of straw. A conflagration ensues in which the burning straw ignites neighbouring pieces of wood; in this fashion the entire barn is consumed in flames.

This example illustrates Mackie's concept of the INUS condition and the full (or 'philosophical') cause. Recall that an INUS condition is a cause in ordinary usage. In this case, the dropping of the burning cigarette is the cause of the barn burning down. There is no question that the selection of one INUS condition or another is highly pragmatic and context dependent. In some cases, such as a spark from a workman's torch, the presence of combustible material might be singled out as the cause. The full cause is a disjunction involving such terms as the dropping of a burning cigarette, a spark from workman's torch, being struck by lightning, spontaneous combustion of hay stored in the barn, and a deliberate act of arson. Recall also that each of the disjuncts contains enough other items to constitute a sufficient condition for the effect. This complex formula states a necessary and sufficient condition for the effect, that is, the burning of the barn.

Mackie's account requires two comments. First, in many cases, and this is no exception, we are not given complete sets of disjuncts. For instance, burning material blown onto the roof from a nearby forest fire was not mentioned. As we have seen, Mackie explicitly admits that the formulations of full causes typically contain '*elliptical* or *gappy* universal propositions' (Mackie 1974, p. 66); the gaps are the reflection of sufficient conditions that are unknown at any particular time. It is difficult to see how such propositions can constitute objective descriptions of full causes; they are epistemically relativized to the knowledge situations of the persons involved. Second, as also previously noted, Mackie explicitly maintains that the statement of the full cause formulates conditions both necessary and sufficient for the effect against the background of a causal field. Items in the field – standing conditions – cannot qualify as causes. Among other things, the field might cover such conditions as the presence of oxygen. Again, there are special circumstances – a particular laboratory experiment, for example – in which the presence of oxygen would be singled out as the cause. It seems obvious that the selection of the causal field is guided by pragmatic considerations, and is, therefore, context dependent.

In view of the foregoing examples, as well as many others, I conclude that cause–effect statements are almost always – if not always – context dependent. (It is easy to find arguments to support context dependency; the hard part is to locate the context independent factors.) If this conclusion is correct, it means that most cause–effect statements lack full objectivity. It seems clear that Mackie has not succeeded in finding causation 'in the objects'. Given this conclusion, it appears that we face a serious problem in the attempt to provide a realistic account of causality. Nevertheless, I think it can be done, but only by going to a different level.

Complete causal structure

The concept of a direct cause–effect relation involves many subtleties. In order to facilitate its further clarification, I shall introduce the concept of a complete causal structure. Having the concepts of causal process, causal transmission, and causal interaction at our disposal, we can do so quite straightforwardly. The complete causal structure of any convex chunk of space-time – that is, of the universe – is given by the entire network of causal processes and causal interactions contained in this selected region. It must include an account of the conserved quantities transmitted by the processes and of those exchanged in the interactions. Assuming that we are dealing with parts smaller than the entire universe, we will have to take account of the processes entering or leaving that section, and with the conserved quantities they bring in or take out.

My notion of a complete causal structure is closely related to ideas of Peter Railton (1981) and Christopher Hitchcock (1993). In his discussions of scientific explanation, Railton formulated the concept of an 'ideal explanatory text', which would include all facts, no matter how insignificant they might seem, that are in any way relevant to the explanandum under consideration. In the case of causal explanation, the ideal explanatory text would be essentially the same as the complete causal structure. The fundamental difference between them is that Railton's construction is a text (a linguistic entity) whereas my complete causal structure is a complex physical entity. Railton stated emphatically that seldom – if ever – do we attempt to write out the ideal text; rather, we seek to illuminate particular parts in which we happen to be interested. In order to do so, we attempt to furnish explanatory information, which involves some part or aspect of the ideal text. Selection of the part to be illuminated is clearly a pragmatic matter, and different investigators will choose different aspects to examine and different levels of detail.

In his work on probabilistic causality, Hitchcock regards probabilistic causal statements as revealing features of particular interest in an underlying probability space. This underlying probability space is, of course, an abstract mathematical construction, but, like many other mathematical abstractions, it is applied to the physical world. Again, the decision regarding the parts to which causal language is to be applied is a highly pragmatic matter.

One basic characteristic shared by Railton's ideal text, Hitchcock's probability space, and my complete causal structure is their objectivity. They are correct or incorrect without regard to context or other pragmatic considerations. This is a fundamental aspect of my claim to furnish a realistic account of causality. The complete causal structure is a fact of nature that exists quite independently of our knowledge or interests; it is not epistemically relativized. It is an extremely complex entity, but that is because the world is extremely complex. Statements about the relations between causes and effects are usually highly selective, and they are typically context dependent. For example, when we discuss Kim's breaking of the window in the course of the baseball game, we ignore the many collisions of the ball with molecules in the air; it is sufficient to take account of the momentum of the ball after it has interacted with the bat. The collisions with the molecules are part of the complete causal structure, but they are not germane to the story.

From the outset, Mackie's goal was to find causation 'in the objects' – that is, to provide a fully objective, non-mind-dependent, account of causation (1974, pp. 1–2). His effort produced three items: (1) the concept of causal priority, specified without reference to temporal priority, (2) the concept of an INUS condition, and (3) the concept of the full cause. I have shown, in connection with the pinball example, that his analysis of causal priority in terms of causal fixity is entirely untenable. However, Reichenbach's conjunctive forks might be imported to determine causal priority. That is the method I am adopting in this chapter.

The concept of an INUS condition, it seems to me, provides a useful tool for the analysis of cause–effect relations, but it is doubtful that such causes can be fully objective and context independent. Mackie was entirely clear on this point. He hoped to find objectivity in the full cause, but, as already noted, his invocation of '*elliptical* or *gappy* universal propositions' (1974, p. 66) gives the strong appearance of epistemic relativity or context dependence. In the pages that follow immediately (pp. 67–75), he makes a good case for the practical utility of such universals; this shows that they have significant pragmatic virtues, but it does not make a convincing case for full objectivity.

Throughout *The Cement of the Universe*, Mackie makes many casual references – more or less in passing – to causal processes, causal mechanisms, interactions, and common-cause configurations.[29] What he failed to see, in my opinion, is the fact that these are precisely the kinds of entities that form the objective basis for our causal claims. Moreover, they manifest another of Mackie's main desiderata, namely, the capacity to distinguish between causal and non-causal sequences of events (p. 29). In the foregoing discussion, I have taken care to show how we can make objective distinctions between causal processes and pseudo-processes and between causal interactions and mere spatio-temporal intersections. My conclusion is that Mackie conducted his analysis on a relatively superficial level, where pragmatic considerations and context dependence play legitimate roles, but that he virtually ignored the objective non-contextual grounding which underlies his level of analysis.

I remarked above that there is great latitude in deciding what constitutes a single process or single interaction. To the astrophysicist, I suggested, Earth, travelling in its orbit, is a single process for most purposes.[30] To the geophysicist, the history of our planet is a complex set of interactions and processes. To a casual observer, a meeting of two people might be a single interaction; to those involved, the meeting might have considerable social and psychological complexity. In the complete causal structure, all of the details are present, and the complexity of some processes that we often regard as single processes is exhibited. In fact, the complete causal structure reveals the extent to which it is permissible to treat complex processes and interactions as simple entities for various purposes.

As I see it, the complete causal structure is limited to situations that do not involve quantum phenomena, where it is well known that causal stories encounter severe difficulties. Below the complete causal structure, there is, so to speak, a quantum mechanical substructure. I do not profess to understand this substructure, beyond being strongly convinced that quantum mechanisms do not conform to the specifications of normal causality. This happens when the wave–particle duality of light or matter enters the context in any significant way.

Cause–effect relations

I do not think it is profitable to try to define such terms as 'cause' and 'effect' in any precise way. They are part of the common idiom; they are used quite loosely and are highly context dependent. Nevertheless, statements about cause–effect relations furnish valuable objective information about the world with respect to our wide range of purposes, interests, and background knowledge. I cannot give an exhaustive list of such applications, but a few examples should convey the main point.

The most obvious context is a situation in which we hope to exercise control by furnishing causes that bring about desirable results or by eliminating conditions that lead to undesirable situations. We search for the causes of airplane crashes and for the means to prevent or cure diseases. The common-cause configuration is crucial in ascertaining what measures should be taken to overcome medical, psychological, or social problems. Treating the symptom rather than the disease frustrates attempts at amelioration. Causal knowledge of this sort is so much a part of daily life that some philosophers (for example, Gasking 1955, von Wright 1971) have tried to characterize causality in terms of manipulability. While manipulability is undeniably an important aspect of causality, I agree with Mackie in viewing a general manipulability account as excessively anthropocentric.

Cause–effect relations figure prominently in the assignment of moral or legal responsibility. My only experience as a juror was on a case in which spoiled food was the alleged cause of the illness of several members of a family. The case was terminated by a settlement between the parties outside of the courtroom. This outcome was somewhat frustrating because the terms of the settlement are confidential. It seemed to me to be an almost perfect real-life case for application of Mill's methods. Readers of the popular novels of Patricia Cornwell – whose fictitious female protagonist Kay Scarpetta has the role of Chief Medical Examiner in Virginia (USA) – learn a great deal about the importance of ascertaining the cause of death when murder is involved.

Cause–effect relations pertain crucially to the transmission of information. If the sender is human, this is just one form of manipulation. We understand how to produce and send messages. However, the message need not be the result of a voluntary act of any intelligent being.

Astronomers who analyse the spectra of light from celestial sources receive information about the chemical constitution of these bodies.

The objective basis for the use of cause–effect relations is that, no matter how complex the case, it is fundamentally reducible to a network of causal processes and interactions. Usually, however, much of the complexity is dispensable. Even such an ordinary act as starting your car is quite complex. After the key has been inserted into the ignition switch, your hand turns the key, which results in the closure of an electrical circuit that, in turn, permits an electric current to flow from the battery to the starter (an electric motor) and to the ignition system that yields sparks to ignite the fuel in the cylinders. In addition, the fuel injector must supply the fuel, and the sparks and fuel injections must be timed in a proper sequence. Even this description presupposes a rich causal field of background conditions. This combination of processes and interactions is embedded in a real – and much more complex – network of processes and interactions. One of the most significant uses of our knowledge of cause–effect relations is to separate useful from useless information in a given context.

This kind of separation enables us to make practical use of such techniques as Mill's methods and controlled experiments to investigate cause–effect relations. It enables us to treat complex combinations of causal interactions and processes as single processes in appropriate contexts. It enables us to treat large assemblages, such as the gas in a container, as a single unit for some purposes; for example, heating causes an increase of pressure in a container of fixed volume. The cause–effect approach often enables us to avoid innumerable useless details. These are pragmatic virtues; it is no surprise that they are context dependent.

In a different context, such as the study of Brownian movement, the individual interactions between small numbers of molecules and individual pollen particles are essential. This illustrates a common research strategy, namely, to look at phenomena on a smaller scale when the larger scale view is unsuccessful. The details are present in the complete causal structure; they can be exposed when it is advantageous to do so. This particular example had crucial importance, around the turn of the twentieth century, in the debate between energeticists and kinetic theorists in thermodynamics.

Conclusion

At the beginning of this chapter, I stated the aim of providing a realistic account of causation in the sense that causal relations should have the same status as ordinary medium-sized material objects. Obviously, we observe such things as billiard balls, baseballs, bats, windows, and children. These are causal processes. There are, of course, causal processes, such as electromagnetic radiation in the radio and television frequency range, that we cannot observe directly. There are also ordinary material objects, such as viruses and molecules, that we cannot observe directly. We observe such causal interactions as collisions of billiard balls and shattering of windows. There are also collisions of individual molecules that we cannot observe directly. Causal processes and interactions thus seem to be on a par with ordinary material objects – both observed and unobserved.

It should be noted that, in observing objects and events of the types just mentioned, we may not be observing them as causal processes or as causal interactions. Nevertheless, the discussion has amply shown, I believe, the kinds of empirical experiments that can be performed to ascertain whether a process is causal or an intersection is an interaction. Performing such experiments does not violate the principles of Humean empiricism. Marking techniques, Mill's methods, and controlled experiments have already been mentioned. The behaviour of conserved quantities plays a key role in my analysis; however, tracing the trajectories of these quantities is not required. In fact, it is not necessarily an efficient method for discovering or demonstrating causal relations.[31]

Although my 'official analysis' appeals to conserved quantities at the fundamental level, I do not mean to suggest that the richness and complexity of the world are exhausted in these properties. I have taken causal transmission to require that some conserved quantity or other be present in a causal process, but causal processes transmit many other kinds of properties as well. Hume's billiard balls transmit colour, solidity, elasticity, and shape. They transmit blue chalk marks that come from the cue stick. Baseballs transmit standard patterns of stitching in their covers and the trademarks of their manufacturers. Bullets transmit characteristic marks of the guns from which they were fired.

Causal processes are, after all, the channels of communication that we find in the physical world quite apart from human thoughts, intentions, and desires. Electromagnetic radiation exists in nature; we have learned to use it to transmit information, music, entertainment, and commercial messages. We have learned to make paper, to write on it, and to convey our thoughts and feelings. The term 'communication' is not meant to include only transmission of information by intentional acts. Fossils inform us of types of organisms that lived on Earth at much earlier times.

There is one final caveat. In no sense have I been attempting to provide a conceptual analysis of causation that would apply to all possible worlds; I have no idea what such a thing would be or how it could be done. Indeed, I do not mean to suggest that causation, as here characterized, applies to all domains in our actual world. I do not believe that what might be called 'normal causation' applies in the domain of quantum mechanics – the domain in which wave–particle duality manifests itself.[32]

Moreover, I make no claims to possess an account of mental causation, if there is such a thing. I have speculated – and it is no more than speculation – that all of the causal processes and interactions related to our conscious experience occur in the brain, and that consciousness is a collection of pseudo-processes. This form of epiphenomenalism makes consciousness an analogue of what we see on the screen of the cinema or television set. The main problem is that I have no idea what constitutes the 'screen' during conscious awareness.

I realize that the theory I am proposing has a highly reductionistic flavour. It seems to me that my account should hold in the natural sciences – including biology, but not quantum mechanics. I am not confident that it is suitable for psychology and the social sciences. In the preceding paragraph, I mentioned my reservations with respect to psychological phenomena. Where interpersonal relations are concerned, I would claim only that causal processes, transmission, and interaction are necessary for recognition of other people and communication among them. The importance of such causal mechanisms should not be overlooked. Words, spoken or written, can break someone's heart, gladden someone's day, or incite someone to violence. Printed messages can amuse, educate, or mislead. Whether other kinds of causation are involved in social intercourse I leave as an open question for philosophers of psychology and of the social sciences.

My aim has been to examine causality at what might be characterized as the 'deepest metaphysical level'. The account that has emerged removes this concept from the field of metaphysics and transports it to physics. If this goal has actually been achieved, I count it as philosophical progress.

Notes

1　This is one of Mackie's examples (p. 44). Although he introduced it for a different purpose, it serves well as an illustration of INUS conditions. The further elaboration of this example is mine.
2　Mackie attended the conference that year, but he was not with us in the bar that night.
3　Sosa and Tooley (1993), is an anthology in which many other approaches are represented.

4   The concept of a process is similar to that of a causal line in Russell (1948). Russell, however, failed to distinguish between causal lines and pseudo-causal lines.

5   A material object at rest in our frame of reference will be in motion with respect to other reference frames, but that is not a crucial consideration. The important factor is its endurance through a stretch of time.

6   Tidal forces might, in some cases, be relevant.

7   The poster was a gift from Michele Marsonet.

8   Light travels at different speeds in various media, but never at a speed greater than its speed in a vacuum.

9   There has been speculation about 'tachyons' — particles that always travel at speeds greater than light — but, to the best of my knowledge, there is no evidence whatever that such particles actually exist.

10 People often say that, according to the theory of relativity, nothing can travel faster than light. This is not true unless we specify clearly that pseudo-processes are not 'things' for purposes of our discussion.

11 Since the speed of light is roughly a billion (in British English, a thousand million) feet per second (that is, one foot per nanosecond), when the circumference exceeds a billion feet, the spot will be travelling faster than light. That translates into a circle with a radius of a bit more than 30,000 miles (much less than the distance from Earth to the moon).

12 Most pulsars radiate at radio frequencies; the Crab pulsar is unusual in this respect.

13 When we consider causal interactions, we will see that causal processes can sometimes intersect without causing lasting changes, for example, the intersections of light waves.

14 The same is true of actions seen on television screens. It is only the means of projection that differ.

15 This modification of the white light as it passes through the film is a complex example of the kind of marking we discussed in connection with the light from La Lanterna.

16 Notice that Zeno appears to be committing the elementary fallacy of composition. This is a nice example; it is hard to find non-trivial cases of such basic fallacies.

17 To maintain a correspondence between points of space and moments of time, we might take the centre of mass of the arrow as the representative point.

18 What happens to the ball originally in motion after the collision depends sensitively on its spin prior to the collision. Hume did not provide enough details to enable us to know the outcome with respect to the first ball.

19 Of course, the egg develops inside the hen before it is laid, so perhaps there is some ambiguity as to when the single process splits into two. For present purposes, it is sufficient to take the exit of the egg from the hen's body as the point at which the single process becomes two.

20 For a time after the mouse is ingested, it can be said to exist as a separate process inside of the snake, but for purposes of illustration it will be satisfactory to consider the snake and mouse merged as soon as the mouse is entirely within the body of the snake.

21 It would be nice to think that the whole range of interactions could be reduced to Feynman diagrams, but the problems of applying causal concepts to quantum mechanics may be utterly insuperable. At any rate, we are far from realizing this pleasant dream.

22 When speaking of interactions, the word 'causal' is otiose; all intersections that qualify as interactions are causal.

23 See Salmon (1997, pp. 463–4), for a discussion of the italicized phrase.

24 I have purposely left the direction of transmission ambiguous; it can be supplied by means of conjunctive forks in a manner I have explained in Salmon (1981).

25 Dowe, in recent personal correspondence, pointed out that I have never answered this question in any published work. I was shocked by the realization that he was absolutely correct. The remainder of this chapter offers an attempt to fill this gap. The problem is much more profound and difficult than I had previously imagined.

26 According to special relativity, there can be no direct causal connection between pairs of events that have a space-like separation.

27 The term 'normal' is meant to exclude such sounds as sonic booms.

28 I could go on mentioning further relevant details – for example, the interactions of the balls with the cushions and the amount of linear momentum initially imparted to the cue-ball – but the degree of context dependence in this example is already well illustrated.

29 The end of ch. 3, pp. 82–7, is one of many rich sources.

30 As mentioned in note 6, the tidal forces might be relevant in some contexts.

31 The results of such experiments are, of course, fallible. Thus, the kinds of counterexamples that had been brought against Reichenbach's analysis in terms of transmission of marks do not count against the use of marking techniques in the detection of causal processes.

32 I realize that baseballs have a wave aspect, but it has no empirical significance for such massive objects.

References

Carnap, Rudolf (1950), *Logical Foundations of Probability*, Chicago, University of Chicago Press.

Dowe, Phil (1992), 'Wesley Salmon's Process Theory of Causality and the Conserved Quantity Theory', *Philosophy of Science*, 59, pp. 195–216.

Dowe, Phil (2000), *Physical Causation*, Cambridge, Cambridge University Press.

Fales, Evan (1990), *Causation and Universals*, New York and London, Routledge.

Gasking, Douglas (1955), 'Causation and Recipes', *Mind*, 64, pp. 479–87.

Hitchcock, Christopher Read (1993), 'A Generalized Probabilistic Theory of Causal Relevance', *Synthese*, 97, pp. 335–64.

Hume, David (1888) [1739–40], *A Treatise of Human Nature*, L. A. Selby-Bigge (ed.), Oxford, Clarendon Press.

Hume, David (1955) [1740], 'Abstract of *A Treatise of Human Nature*', in C. W. Hendel (ed.).

Hume, David (1955) [1748], *An Inquiry Concerning Human Understanding*, C. W. Hendel (ed.), Indianapolis, Bobbs-Merrill. [Modern spelling of 'Enquiry' introduced by the editor.]

Hume, David (1955) [1776], 'My Own Life', C. W. Hendel (ed.), Indianapolis, Bobbs-Merrill.

Keynes, J. M. (1952), *A Treatise on Probability*, London, Macmillan and Co.

Mackie, J. L. (1974), *The Cement of the Universe*, Oxford, Clarendon Press.

Mill, J. S. (1874), *A System of Logic, Ratiocinative and Inductive*, New York, Harper and Brothers, 8th ed.

Railton, Peter (1981), 'Probability, Explanation, and Information', *Synthese*, 48, pp. 233–56.

Reichenbach, Hans (1956) [1928], *The Direction of Time*, Berkeley/Los Angeles, University of California Press.

Reichenbach, Hans (1958) [1928], *The Philosophy of Space and Time*, New York, Dover. Original German ed., 1928.

Russell, Bertrand (1922), 'The Problem of Infinity Considered Historically', in *Our Knowledge of the External World*, London, George Allen and Unwin, pp. 159–88.

Russell, Bertrand (1948), *Human Knowledge: Its Scope and Limits*, New York, Simon and Schuster.

Salmon, Wesley C. (1967), 'Carnap's Inductive Logic', *Journal of Philosophy*, 64, pp. 725–39.

Salmon, Wesley C. (1969), 'Partial Entailment as a Basis for Inductive Logic', in *Essays in Honor of Carl G. Hempel*, Nicholas Rescher (ed.), Dordrecht, D. Reidel, pp. 47–82.

Salmon, Wesley C. (1981), 'Causality: Production and Propagation', in *PSA 1980*, P. D. Asquith and R. N. Giere (eds), East Lansing, Michigan, Philosophy of Science Association, pp. 49–69. Reprinted in Salmon, 1998, pp. 285–301.

Salmon, Wesley C. (1994), 'Causality Without Counterfactuals', *Philosophy of Science*, 61, pp. 297–312. Reprinted in Salmon, 1998, pp. 248–60.

Salmon, Wesley C. (1997), 'Causality and Explanation: A Reply to Two Critiques', *Philosophy of Science*, 64, pp. 461–77.

Salmon, Wesley C. (1998), *Causality and Explanation*, New York, Oxford University Press.

Sosa, Ernest and Michael Tooley (eds) (1993), *Causation*, Oxford, Oxford University Press.

von Wright, G. H. (1971), *Explanation and Understanding*, Ithaca, NY, Cornell University Press.

Whitehead, A. N. and Bertrand Russell (1910–13), *Principia Mathematica*, 3 vols, Cambridge, Cambridge University Press.

Chapter Seven

Realism and Anti-Realism from an Epistemological Point of View

Paolo Parrini

A preliminary terminological clarification

First of all, I would like to say that, since this chapter is written in English, I have decided, starting from the title, to use the term 'epistemological' instead of 'gnoseologico' as I should have done if I were writing in Italian. In English the words 'epistemology' and 'epistemological' indicate what in Italian is called 'filosofia della conoscenza' ('philosophy of knowledge') or 'gnoseologia' ('gnoseology'), whereas in Italian the terms 'epistemologia' and 'epistemologico' indicate something which is nearer to the philosophy of science. Nevertheless, some English-speaking authors use the words 'epistemology' and 'epistemological' to indicate not only philosophy of knowledge in general, but also that part of philosophy of science which deals with the problem of the validity of scientific claims. For instance, David Papineau (1996, p. 290) begins his paper 'Philosophy of Science', written for *The Blackwell Companion to Philosophy*, by saying that

> philosophy of science can usefully be divided into two broad areas. On the one hand is the epistemology of science, which deals with issues relating to the justification of claims to scientific knowledge. ... On the other hand are topics in the metaphysics of science, topics relating to philosophically puzzling features of the natural world described by science.

In this chapter I will use the words 'epistemology' and 'epistemological' to refer in a general way to the questions concerning the validity of our cognitive claims, both the ones put forward in our everyday life and the ones put forward in science. And it is from this point of view – and not from others such as, for instance, a logical, ontological, semantic, or intentional point of view – that I will discuss the contrast between realism and anti-realism.

The contrast between realism and anti-realism

When we look at the contrast between realism and anti-realism both in a wide historical perspective and from the point of view of the recent epistemological debate, it seems to me that no decisive arguments have been offered either in favour of various forms of realism or in favour of various forms of anti-realism (phenomenalism, instrumentalism, logical idealism, and so on). Both conceptions have advantages and defects which somehow balance each other out.

To explain what I mean, in 2/A–R and in 2/R I will list three main anti-realist theses and three main realist theses which have been all well argued in the epistemological literature on this topic.

2/A–R. *Three main anti-realist theses*

In order to present the question more clearly, I will take as my starting point the ideas put forward by the supporters of the anti-realist position. It seems to me that the anti-realists have been able to support pretty well the following three main theses.

(A–R1) The first thesis may be called the thesis of epistemic relativism. According to it, our knowledge cannot do without subjective epistemic assumptions. As I will show later on, this thesis must not be confused with the thesis of radical relativism. Epistemic relativism maintains that cognitive activity 'is done from within a *Weltanschauung or Lebenswelt*' (Suppe 1977 [1973], p. 126). This approach is connected with all the philosophical tendencies which, following Kant, have emphasized the subjective components of cognitive activity. However, the connection with the pragmatist tradition and historicism sets relativism apart from Kant. The uniqueness of the categorical system is denied, great importance is attributed to the historical change of categories and special attention is paid to the sociological factors which influence the acceptance, formulation, and rejection of world-views.

The core of epistemic relativism is the idea that knowledge (common and scientific) is set within a conceptual scheme more or less uniformly shared by a community of individuals engaged in certain cognitive practices. This epistemic framework is a research perspective which is closely tied to the structure of the language which is used, it incorporates a 'thought style' (Fleck [1935] 1979) concerning a variety of more or less clearly characterized phenomena, delimits the class of legitimate problems and sets the standards of acceptable solutions. Some of these conditions are 'tacit' presuppositions which can be made explicit in the passage from one context of research to another; the new context, in its turn, will be characterized by different tacit factors. There is no reason to believe that there are absolutely fixed and invariant assumptions; however, the change of conceptual schemes requires the more or less explicit adoption of other epistemic assumptions. Certain assumptions constitute the core of the theoretical structure. As such, they represent an epistemologically distinct class of statements, which are neither a priori, since they can be modified according to experience, nor trivially a posteriori, since they are 'protected' from direct empirical refutation.

Epistemic relativism largely prevailed following the attacks of the 'new philosophers of science' (in the Italian terminology), philosophers such as N. R. Hanson, T. S. Kuhn and P. K. Feyerabend, on the standard conception of scientific theories. Actually, we now know that logical empiricists' ideas were much more complex than their critics thought and that in a way they were anticipations of the new philosophy of science. But I do not wish to discuss this question here. My aim is only to underline the dependence of the cognitive enterprise upon a more or less organic complex of different subjective epistemic conditions. The plausibility of this thesis depends upon the existence of a marked gap between experience and the complex of our 'beliefs and convictions'. According to epistemic relativism, it is not possible to bridge this gap without referring to a set of linguistic, theoretical, and methodological-axiological conditions, entailing corresponding forms of epistemic relativism (linguistic relativism, theoretical relativism, and methodological and axiological relativism) (see Parrini 1998 [1995], II/1–3).

As I have already said, the thesis of epistemic relativism must not be confused with radical relativism. Epistemic relativism simply maintains that our cognitive access to reality and truth is linked to some subjective conditions of different kinds (linguistic, theoretical, and methodological-axiological). By contrast, radical relativism maintains that truth and reality themselves (and not our ways of access to truth and reality) are relative to a conceptual scheme. In this perspective the possibility of a plurality of different conceptual schemes implies the existence of a plurality of different truths and worlds. For instance, Sapir (1951, p. 162) endorsed the radically relativist thesis that the 'world in which different societies live are distinct worlds, not merely the same world with different labels attached' – a statement which matches Kuhn's (1996, p. 111) 'exclamation' 'that when paradigms change, the world itself changes with them'.

(A–R2) The second well-argued anti-realist thesis is the following. What we know as object, and call 'object', is in fact the result of a synthesis of experience effected on the basis of different kinds of epistemic assumptions (linguistic, theoretical, and methodological-axiological). This thesis comes mainly from Kant who very clearly put it forward in the *Critique of Pure Reason*, when he gave his famous answer to the question: 'What ... is to be understood when we speak of an object corresponding to, and consequently also distinct from, our knowledge'? (A104f). This is Kant's (1985 [1781, 1787], A104f, p. 134f) answer:

> It is easily seen that this object must be thought only as something in general = x, since outside our knowledge we have nothing which we could set over against this knowledge as corresponding to it. Now we find that our thought of the relation of all knowledge to its object carries with it an element of necessity; the object is viewed as that which prevents our modes of knowledge from being haphazard or arbitrary, and which determines them *a priori* in some definite fashion. For in so far as they are to relate to an object, they must necessarily agree with one another, that is, must possess that unity which constitutes the concept of an object.
>
> But it is clear that, since we have to deal only with the manifold of our representations, and since that x (the object) which corresponds to them is nothing to us – being, as it is, something that has to be distinct from all our representations – the unity which the object makes necessary can be nothing else than the formal unity of consciousness in the synthesis of the manifold of representations.

Later on, this basic Kantian idea would be taken up and developed by Ernst Cassirer. According to Cassirer, before Kant 'the thing and the self had always been projected, in order to be understood in their connection, upon a common metaphysical background'. Kant sought 'the logical, universally valid and fundamental form of experience in general, which must be binding in the same way both for "internal" and for "external" experience'. In this sense, we may apply to Kant's philosophy 'the words of Schiller's famous epigram: it knows nothing about the thing and nothing about the soul'. The object of Kant's inquiry is 'no longer composed of the things, but of the judgements on the things' (Cassirer 1922, p. 662). For Kant, 'judgment and object are strictly correlative concepts, so that in the critical sense, the truth [= reality] of the object is always to be grasped and substantiated only through the truth of the judgment' (Cassirer 1981, p. 285). In critical-transcendental philosophy, 'it is not because there is a world of objects that there is for us, as their impression and image, a world of cognitions and truths'. On the contrary: if there is for us an organization which we can designate as an order not merely of 'impressions and representation', but also of objects, this is because there are 'unconditionally certain judgements – judgements whose validity is dependent neither on the individual empirical subject from which they are formed nor on the particular empirical and temporal conditions under which they are formed' (Cassirer 1981, p. 148).

As Abel Rey said, illustrating the passage from the old positivism to Poincaré's new positivism, it was essential to realize that '[o]bjective experience and mind are functions of each other, imply each other, and exist by virtue of each other'. Through Dummett's mediation, something similar has come about in American philosophy. With Putnam's passage from external to internal realism, at Harvard, the idea 'that objects and reference arise out of discourse rather than being prior to discourse' is being taken into serious consideration again: '[r]eference ... is not something prior to truth; rather, knowing the conditions under which sentences about, say, tables, are true is knowing what "table" refers to (as on a disquotational theory of *reference*)'.

(A–R3) The third thesis well established by anti-realists is the following. There is an apparently indissoluble tie between scepticism and realism. I would like to illustrate this thesis by referring to the thought of Giulio Preti, an Italian philosopher less well known abroad than he deserves. His proposal aspires to connect in a non-trivially eclectic synthesis demands which come from the most significant philosophical experiences of our time: neo-Kantianism and phenomenology, conventionalism and neo-positivism, historicism and pragmatism. His integration was not always respectful of the motivations that lie at the bottom of the different traditions of thought that he brought together. What is relevant here is not the overall validity of his position. Instead, I would like to stress the link he establishes between realism and scepticism, especially in one of his most recent writings on the problem of knowledge. Actually, I think that the strongest feature of this line of thought descending from Kant lies in the fact that it has brought to light the indissoluble tie between metaphysical realism and epistemological scepticism. As Preti has repeatedly stressed, there can be no overcoming of the sceptical stance as long as we remain committed to a formulation of the problem of knowledge tied to the 'dogmatic' acceptance of a knowing subject opposed to the existence *an sich* 'of a real world independent of knowing'. It is this ontologically prejudiced approach which legitimizes the sceptical stance and makes the problem itself into a contradiction or a 'non-sense'.

Apart from the question of contradiction and of 'non-sense', which I will discuss shortly, according to Preti scepticism is nothing but the unfolding of the consequences of the original ontological duplication. In substance, scepticism limits itself to reaffirming that knowing does not contain its own criterion in itself, that is, it does not contain the criterion of truth and error. *This criterion is external to knowing, it is in something else than knowing.*

This is particularly clear if we try to consider the scholastic formula *veritas est adaequatio intellectus et rei* as the criterion of truth: '[t]hought does not contain any criterion to establish the *adaequatio*, because it has its own criterion outside *itself*, in the *res*. Reality must be "seized", "caught", but we can never know "whether we have caught a real thing or a shadow"'. This is why metaphysical realism and the sceptical stance 'are simply two sides of the same coin'.

We can escape scepticism only if we interpret the sceptical *epoché* as primarily referred to all conceptions which make the criterion of knowing or of truth consist in a transcendent object in itself which, in principle, is something different from the cognitive representation itself: '[i]f scepticism is realism's self-criticism, it is also the liberation from the realist presupposition'. In order to overcome the sceptical stance, epistemology must cast away all metaphysical presuppositions, recovering an 'internal' sense of 'truth' as 'agreement of every *determinate* factual representation and/or *determinate* discursive process to internal criteria of knowledge, such as, precisely, consistency, coherence, etc.' (Preti 1974, pp. 3, 7, 12).

I would like to stress the parallelism between these assertions by Preti and the conclusions reached by Hempel in the years spanning the 1980s and the 1990s. In *Limits of a Deductive Construal of the Function of Scientific Theories* (1988), he maintained that '[o]n what there is, or what the world is like, we can make no more reasonable judgment than that based on the best world view, or the best theoretical system, we have so far been able to devise' (Hempel 1988, p. 14). And in *Eino Kaila and Logical Empiricism* (1992), he added that lacking 'reasons', even highly conjectural, which would allow us to evaluate verisimilitude, 'we never know how close we are to the truth at any point in the continuing process of formulating, testing, and adjusting our conjectures about the world'. The analysis of scientific procedures shows unequivocally 'that none of the considerations involved in the critical appraisal and the acceptance or rejection of empirical claims [i.e., empirical adequacy, coherence, large scope, simplicity, and so on] has any bearing whatever on the question whether the claims in question are true or likely to be true' (Hempel 1992, p. 48).

In the light of these alternatives, Hempel reached the conclusion that 'the idea of science as a search for truth, for a true description of "reality", has to give way to an epistemically relativized conception of scientific inquiry as directed towards the construction of ever-changing epistemically optimal world-views' (Hempel 1992, p. 51). But it is clear that from Preti's point of view we are obliged to accept Hempel's conclusion only if we start from the idea – which is typical of metaphysical realism – that reality, objectivity, and truth are something that go beyond the cognitive process, that they are what this process must somehow conform to, although we have no criterion for establishing when we reach or do not reach an *adaequatio* however partial or approximate it may be (verisimilitude).

Arthur Fine and Larry Laudan (1984 [1981]) made this point very clearly when they criticized several forms of convergent epistemic realism based on the so-called 'best explanation'. In order to defend forms of realism of this kind one should be able to find a considerable number of inductions capable of connecting the observations of the predictive successes of science with the referentiality and the at least approximate truth of science's posits and descriptions. What the anti-realist stresses – in my opinion successfully – is the impossibility of observing a connection of this kind, since we are not able to say what reality and truth look like independently of our theories. Therefore, we do not possess observations which could constitute the starting point for an inductive inference. As Arthur Fine (1984, p. 85f) says, we are again faced with the 'well-known idea that realism commits one to an unverifiable correspondence with the world'. And, as I already stressed, this very idea is the main argument at the basis of sceptical doubts and of Kant's analysis of the object of representations, which I considered above in A–R2.[1]

2/R. Three main realist theses

To the three main anti-realist theses listed in 2/A–R correspond three alternative theses put forward by the supporters of realism. These three theses can be summarized in the following way.

(R1) The first thesis maintains that the notion of object of knowledge – even if it is meant in an absolute, metaphysical sense – is neither meaningless nor internally incoherent. This thesis is not trivial, because it goes against some claims that anti-realists repeatedly have tried to prove. Representatives of the Kantian and neo-Kantian tradition of thought and of the phenomenological-Husserlian one tried to show not only that metaphysical realism makes knowledge impossible in principle (scepticism), but even more drastically that the notion of reality in itself or *Ding an sich*, on which metaphysical realism is grounded, is a senseless or internally contradictory notion.

For instance, this last thesis was put forward by Wilhelm Schuppe. When proposing his philosophy of immanence (fully criticized by Moritz Schlick in *Allgemeine Erkenntnislehre*), he said that '[a] thought that is directed to a thing makes this thing something thought; consequently, the thought of a thing that is not thought is an unthinkable thought' (see Schlick, [1918, 1925], 1974, p. 196f).

As far as the charge of senselessness made to the idea of transcendent metaphysical reality is concerned, it clearly appeared in some ways of interpreting the Kantian conception of knowledge and the known object. An example of this can be found in the passages by Preti quoted above when illustrating the thesis A–R3 of the anti-realist position. Another particularly meaningful example can be found in what Robert P. Wolff maintained in his well-known 1963 book *Kant's Theory of Mental Activity: A Commentary on the Transcendental Analytic of the 'Critique of Pure Reason'*. According to Wolff, Kant's redefinition of objectivity through the notions of the universal and necessary validity of judgements will seem to many as a way not so much of solving as of evading the Cartesian problem of the agreement between the sequences of our subjective ideas and the world of things. According to them, '[u]niversality and necessity may indeed be among the marks of knowledge, but there is also the belief in an object "out there" (*ob-ject*), standing over against (*gegen-stand*) the subject. If that has been lost, then the result is scepticism, no matter what one calls it'.

Wolff is willing to admit that '[t]his criticism' expresses the point of view of the metaphysical realist', and that it is 'extremely hard' to withstand. Nevertheless, he considers it 'totally without force', because '[t]o some extent it is based on a misunderstanding of Kant's position, a misunderstanding much like that which Dr Johnson manifested' when he thought he was showing the reality of a stone by kicking it. According to Wolff:

> Kant's own statements to the contrary notwithstanding, it is not the teaching of the *Critique* that phenomena are merely in the head, or that in any ordinary sense material objects are not real. But to the more sophisticated objection – that the 'ordinary sense' includes and must include the idea of an ontologically independent object – there can be no answer beyond a careful reiteration of all the reasons why such a demand is self-contradictory. Universality and necessity are all you can get, Kant says in effect. Therefore, they are all it has ever been legitimate to demand. Anyone who persists in asking what the world is *really* like – in other words, who wishes to know what an object is like independently of the conditions of his knowing what it

is like – must then simply be dismissed as not serious. In the terminology of a later philosophical school, he needs to be cured, not answered. (Wolff 1963, p. 322f)

The claim that the critical-transcendental turn shows that metaphysical realism is a nonsense, or a linguistic cramp in need of a cure, seems definitely excessive. It is Kant's very perspective that preserves a meaningful opposition between reality in itself and phenomenal reality. Therefore, there is no reason to dismiss as 'not serious', 'in need of cure', or 'not making sense' anyone who believes that our cognitive claims are true only when they agree with the properties of things in themselves, whether or not the criterion with which to establish the agreement is in principle available to us. The realist metaphysician's claim that Kantian epistemology is a form of idealism, incapable of wholly freeing the known object from its dependence upon the knowing subject, is exactly as 'meaningful'. The proof that in principle we cannot possess a truth criterion which will satisfy the requirements of metaphysical realism and that, therefore, epistemological correspondentism entails scepticism, cannot be considered as a proof of the meaninglessness or of the linguistically pathological character of those requirements.

I would like to add, as a further example, that recently the idea of the senselessness of metaphysical realism has been criticized with reference to Hilary Putnam's abandonment of 'external realism' in favour of 'internal realism'. According to internal realism, the question

> *what objects does the world consist of?* is a question that it only makes sense [emphasis added] to ask within a theory or description. 'Truth', in an internalist view, is some sort of (idealized) rational acceptability – some sort of ideal coherence of our beliefs with each other and with our experiences *as those experiences are themselves represented in our belief system.* (Putnam 1982, pp. 49f, 52)

But many philosophers of science and philosophers of language have continued to hold out against this position of metaphysical realism and scientific realism, justifying this choice by means of arguments connected with the causal theory of reference and with the conception of names as rigid designators.

For instance, Ian Hacking, taking his position precisely with reference to Putnam's claim quoted in the preceding paragraph, declared himself willing to accept, at least at the metaphorical level, the idea that we 'cut up the world into objects when we introduce one or another scheme of description'. But he also declares that he cannot endorse the previous statement that '"Objects" do not exist independently of conceptual schemes'. The 'Inuit are said to distinguish ever so many kinds of snow that look pretty much the same to us', by means of which '[t]hey cut up the frozen North by introducing a scheme of description'. But, according to Hacking, '[i]t in no way follows that there are not 22 distinct mind-independent kinds of snow, precisely those distinguished by the Inuit'. Furthermore, since the Inuit do not ski, it can very well be the case that 'powder snow, corn snow, or Sierra cement spoken of by some skiers neither contain nor are contained in any Inuit class of snow'. Nevertheless, we are allowed to presume that 'there is still powder snow *and* all the Inuit kinds of snow', and that such kinds all refer to 'real mind-independent distinctions in a real world'. 'These remarks', says Hacking, 'do not prove that there is powder snow, whether anyone thinks of it or not. They merely observe that the fact that we cut up the world into various possibly incommensurable categories does not in itself imply that all such categories are mind-dependent' (Hacking 1983, p. 93).

The vigour of realism lies precisely in the undeniable possibility of using our linguistic expressions in order to refer directly to the object, its existence and its properties, independently of the possible conditionings by the linguistic, theoretical, and methodological conceptual schemes within which we operate, and sometimes are forced to operate in the absence of conceivable alternatives. We can change the way we call things as we wish, but they will not change their properties because of this. And we can say the same with regard not only to the conventionality of linguistic symbols, but also to all epistemic conditions: we can always act on a so-called 'objectocentric' expressive plane, saying that the object is what it is, or has or does not have certain properties, independently of the linguistic, theoretical, and methodological presuppositions on which our knowledge of the object depends.

As a matter of fact, those very presuppositions will have to be judged as correct or incorrect depending on whether they are or are not consistent with the object. This was exactly the point that underlay Russell's assertion of the objective value of spatial congruence and his criticism of Poincaré's geometrical conventionalism (Russell 1990 [1899], p. 396f):

> It seems to be thought that, since measurement is required to *discover* equality and inequality, there cannot *be* equality or inequality without measurement. The true inference is exactly the opposite. What can be discovered by an operation must exist apart from the operation: America existed before Christopher Columbus, and two quantities of the same species must *be* equal or unequal before measurement. Any method of measurement is right or wrong

according as it brings out the right or wrong result. M. Poincaré, on the contrary, holds that measurement *creates* equality and inequality.

(R2) The second thesis established with good arguments by the supporters of realism says that the structure and the development of knowledge depend on a given which is not completely determinable by the knowing subject, that is by the epistemic structures of reference – and this dependence suggests quite naturally the idea of a reality independent of knowledge and which knowledge itself strives to reach.

In this case also the thesis has been established both through the analysis of problems posited by traditional philosophies such as the Kantian one and by conceptions recently put forward by the new philosophers of science. Looking at the question from this second point of view – the point of view of the new philosophy of science – it was pointed out that not even the most radical champions of the incommensurability thesis and of the theory-ladenness of the observation were able to deny any role to experience in the process of knowledge, that is were able to deny the well-proved fact that we hold testable empirical expectations which are often frustrated (the well-known stubbornness of data). With respect to staunch empiricists, authors such as Kuhn, Lakatos, and Feyerabend have certainly aimed at attenuating the import of experience in the evaluation of theories; but they were not able, nor willing, to deny it completely. They have opted for a position which on the whole is in line with the outcome of Duhem's criticism of crucial experiments, which endorses the empirical falsificability and confirmability of theoretical systems globally considered.

Kuhn's position constitutes a particularly instructive example. He insists on the idea that cognitive activity presupposes the interpretation of nature by means of a set of 'conceptual boxes'. On the other hand, he also maintains that 'nature cannot be forced into an arbitrary set of conceptual boxes. On the contrary, the history of proto-science shows that normal science is possible only with very special boxes, and the history of developed science shows that nature will not indefinitely be confined in any set which scientists have constructed so far'. If this were not the case, Kuhn would not have been able to talk about the empirical 'anomalies' which affect paradigms (understood as 'disciplinary matrixes') (Kuhn 1987, p. 263).

The Duhemian spirit of this aspect of the question is well condensed in a neglected remark by Lakatos: '[i]t is not that we propose a theory and Nature may shout *no*; rather, we propose a maze of theories, and Nature may shout *inconsistent*'. Nature can, that is, induce us to uphold 'a "factual" statement couched in the light of one of the theories involved, which we claim Nature had uttered and which, if added to our proposed theories, yields an *inconsistent system*' (Lakatos 1987, p. 130; capital letters replaced by italics). Feyerabend himself has explicitly recognized the possibility of refuting 'incommensurable theories ... by reference to their own respective kinds of experience' – though he specifies parenthetically that 'in the absence of commensurable alternatives these refutations are quite weak' (Feyerabend 1987, p. 227).

But it is even more interesting to look at the question from the first point of view, that is from the point of view of the problems raised by the logical or formal idealism of a Kantian and neo-Kantian kind. In this case the question is directly connected to the problem of the *Ding an sich* and thus to the problem of metaphysical realism. In this perspective, we can maintain that in the philosophical tradition good arguments were given for the claim that the critical point of view fails not only to show the meaninglessness of metaphysical realism (see R2 above), but even its emptiness. This aspect of the question is interwoven with the problem posited by the Kantian dichotomy between form and matter of knowledge. It has a history of its own in Kantian literature and can ultimately be traced back to the question of 'determined' empirical knowledge.

Kant poses the question of the possibility of synthetic a priori knowledge, and through it of knowledge in general; but it is clear that an adequate solution of the latter problem cannot be separated from an explanation of the possibility of determined empirical knowledge. Instead – as Broad pointed out – one of the most surprising aspects of the *Critique* and of the *Prolegomena* is that there is never any question about 'what it is that determines the particular shape, size, and position which a particular object is perceived to have on a particular occasion' (Broad 1978, p. 25). The problem has been raised also by other authors (for example, C. I. Lewis), especially in order to show that subjectivist idealism is not capable of accounting for the particularity of experience. It is no coincidence that an incisive formulation of the question can be found already in Herbart's philosophy, where it is connected with the distinction between phenomena and things in themselves (see Parrini 1994a [1990], pp. 210–19).

Herbart, raising the problem of determinate knowledge, had already observed that there must be some relation between things in themselves, the matter of knowledge, and the properties which we attribute to phenomenal objects on the basis of experience. And in this century, the discussion of this aspect of Kantianism has intertwined itself with the vicissitudes of analytic philosophy. The theoretically relevant aspect of the matter was well formulated by Schlick, but it was noted also by Russell and Wittgenstein:

> if the 'phenomena' are appearances of something else, then the mere fact that this 'something else' is that particular reality of which that particular phenomenon is the appearance – this fact enables us to describe the reality just as completely as the appearance of it. The description of the appearance is, at the same time, a description of that which appears. The phenomenon can be called an appearance of some reality only in so far as there is some correspondence between them, they must have the same multiplicity; to every diversity in the phenomenon there must be a corresponding diversity in the appearing things, otherwise the particular diversity would not form part of the phenomenon qua phenomenon, nothing would 'appear' in it. But if this is so, it means that the 'appearance' and the 'appearing reality' have identically the same structure. The two could be different only in content, and as content cannot possibly occur in any description, we conclude that *everything* which can be asserted of the one, must be true for the other also. The distinction between appearance and reality collapses, there is no sense in it. (Schlick 1979 [1932], p. 359]

Of course, if a supporter of Kant's thought was faced with Schlick's attempt to abolish the distinction between phenomenal appearance and reality-in-itself, he could point out that Schlick's attempt depends on an outmoded form of the principle of verifiability. One of the principal teachings of the *Critique* remains the demonstration of the inevitable conditioning exercised on the known object by certain forms of knowledge, whereby it will never be possible to know what things in themselves are like. But the huge amount of literature on transcendentalism is there to prove that, even if we place ourselves in a strictly Kantian perspective, we still have to face the problem of explaining somehow the relation between phenomenal appearances and things in themselves. And it is difficult to consider realism's requests as completely empty or unjustified, if we do not possess a theory of meaning powerful enough to warrant realism's meaninglessness – all the more when we are forced to recognize the variability of the presuppositional system. And this leads us to the third thesis put forward with some success by the supporters of realism.

(R3) The third thesis that realists succeeded in defending is the negative claim that anti-realists did not produce good arguments to establish that our epistemic conditions of knowledge, uncovered by the analysis of the cognitive process, could not as well be structural properties of objects in themselves. Also this thesis has a long story behind it. It can be reduced to the so called objection of the 'neglected alternative' that dates back to Friedrich Trendelenburg in the nineteenth century and to Nicolai Hartmann in the twentieth century.

Starting from the years which followed the formulation of transcendental epistemology some authors showed that in a Kantian perspective, metaphysical realism is so far from being 'meaningless' (see R1 above) that it is possible to formulate the realist objection of the so-called 'neglected alternative' precisely by using the conceptual apparatus of formal idealism, and in particular the distinction between empirical objects and things in themselves. This objection contends that Kant's argument for transcendental idealism is vitiated by the unproven presupposition that certain forms of knowledge (space, time, causality, and so on) could only be either structures of the knowing subject, or structures of reality in itself. The neglected alternative is that they could be both these things at once; after all, if things in themselves are unknowable, it is odd to claim that one can know that they do not have the characters of spatiality, temporality, substantiality, causal concatenation, and so on.

I agree with Robert P. Pippin that this objection resists even the refutation recently attempted by Henry E. Allison (see Parrini 1994a [1990], p. 215 and n. 51). And it seems to me that thesis R3 gains further strength if we consider the possibility of an historical evolution of the forms of knowledge under the pressure of experience as the one envisioned in R2 when considering the problem of determined knowledge. Of course, it will always be possible to contest the legitimacy of projecting the formal and material components onto an ontological background as two entities existing in a separate and opposed manner, by recurring to Cassirer's theory that '[m]atter *is* only with reference to form', and that form '*is valid* only in relation to matter'. Analogously, we could follow Cassirer again, in disqualifying, as unacceptable applications of the categories to objects transcending the field of possible experience, the old questions about the status of cognitive conditions: whether they are 'an element of *being*', or of nature, or 'mere constructions of *thought*', or 'the universal forms of expression of our consciousness' (Cassirer 1953 [1910], pp. 309 and 311). But, as I have said, such objections cannot be equated to a proof of the senselessness of metaphysical realism, and not even of its complete emptiness. And if we acknowledge that cognitive presuppositions can be modified under the pressure of conceptual innovation and of the data of experience, the realist tendency to interpret scientific change as the progressive approximation to the knowledge of a transcendent reality will gain momentum. The very fact that in our century the crisis of Kantian conception was followed by a strong revival of realistic tendencies of several kinds, such as Popper's fallibilist realism or convergent epistemological realism, is a clear illustration of my point.

Proposed 'positive' answer to the epistemological contrast between realism and anti-realism

What seems to emerge from the previous section is that there is a kind of balance of the arguments for and against realism that is analogous to the balance already stressed by Duhem in his *Théorie physique: son objet, sa structure* (1904–6, 1914). While discussing the question whether we must grant physical theory only an economical and representational value or also a classificatory and explanatory function endowed with an ontological import, Duhem recognizes that the methodological analysis of physical theories leads to a dilemma in the face of which we can only appeal to Pascal's 'reasons of the heart that "reason does not know"':

> [T]he analysis of the methods by which physical theories are constructed proves to us with complete evidence that these theories cannot be offered as explanations of experimental laws; and, on the other hand, an act of faith, as incapable of being justified by this analysis as of being frustrated by it, assures us that these theories are not a purely artificial system, but a natural classification. And so, we may here apply the profound thought of Pascal: 'We have an impotence to prove, which cannot be conquered by any dogmatism; we have an idea of truth which cannot be conquered by any Pyrrhonian scepticism'. (Duhem 1962, p. 27)

I think that there is something deeply true in Duhem's claim and that the contrast between realism and anti-realism cannot be solved by appealing to logical-analytical argumentation or evidence of an empirical-factual kind. But I also think that, drawing our inspiration from Herbart's conception of philosophy as *Bearbeitung* of our main notions, we can try to give an answer to the question. What I am proposing is a theoretical reconstruction of a 'non-coercive' kind (I employ here the expression of R. Nozick 1981, p. 3f), that is a reconstruction that does not pretend to be founded on absolute certainty and definitiveness, but that aims at competing with rival proposals hoping to be judged as the best one, at least on the whole (see Parrini 1998 [1995], VII/4). In a recent book of mine (1998 [1995]), I have tried to work out a similar proposal starting from the idea that it is this very balance of the arguments in favour of and against realism that lends support to the suspicion that the question, *qua* metaphysical question, is not genuinely scientific, and that, therefore, it is not possible to give it an answer analogous to those that are commonly offered by science.

A similar view had indeed already been held by the neo-positivists (particularly by Carnap) when, relying on the verification principle, they distinguished between 'empirical reality' and 'metaphysical reality', numbering the problem of metaphysical realism among the questions devoid of cognitive meaning. And something analogous has recently come to light in 'minimalist' positions such as Fine's, and in recent criticisms, such as Laudan's, of realist-metaphysical interpretations of the historical development of science. Fine and Laudan have pointed out that the verdict on metaphysically interpreted epistemological contrapositions transcends the conditions of decision proper to the normal scientific procedure: it goes beyond the decision methods normally employed in scientific inquiry.

Certainly, such criticisms do not authorize us to conclude that the metaphysical contrast between scientific realism and instrumentalism, between realism and anti-realism of different kinds, is 'devoid of sense' or even simply 'devoid of cognitive value'. These criticisms, though, do allow us to say that this contrast does not belong to the field of science. As far as we know, science can only posit – in a more or less hypothetical way, and referring chiefly to its means of experimental manipulation (rightly reappraised by Hacking 1983) – a certain domain of entities and give us the description of it which it, science, considers most adequate, distinguishing between things which are real and things which are imaginary, dreamed, and so on. But science has nothing to say about reality in itself, that is, about the absolute correctness of its positions and descriptions. In this typically positivist claim (which Hempel has firmly maintained), it seems that there is a kernel of truth – of positivistic truth – which is difficult to drop, and which has really not been completely abandoned even after the crisis of neo-positivism (see Parrini 1994b).

But, if we adopt this approach, which is completely independent of the verification principle, it is possible to give philosophical value to anti-absolutism only if we take up two very problematic assumptions. Once we have abandoned the verification principle, then we will no longer be able to say – as Carnap and Schlick claimed to do – that metaphysical realism, insofar as it is founded on the idea of a transcendent reality *an sich*, is a thesis devoid of empirical content and therefore of meaning. Consequently, it will be possible to pass from the statement that questions concerning reality in itself do not have scientific value to the wider one that they do not have cognitive value, only if we assume the validity of (1) a scientistic position which ascribes genuine cognitive value to science only; and (2) a relativist conception of truth and cognitive objectivity which ascribes to them an exclusively intratheoretical or intrasystemic value.

In this way, though, we are not only endorsing scientism, a position which is not too seductive for various reasons. We are also running into the big difficulties of radical relativism. If taken in large doses, relativism is a highly problematic philosophical position. It seems very arduous to give up any distinction between what our theories say, the reasons we have for believing them, and their truth and falsehood, in a sense of truth and falsehood not purely relative to a set of epistemic conditions of reference. If we adopt a radically relativist perspective, we get tangled up in two types of problems which make such a position most unattractive.

The first difficulty consists in the fact that radical relativism ends up compromising those aspects of the notion of truth according to which truth is seen as a unitary axiological and normative ideal. This point has always been emphasized not only by metaphysical realists, but also by philosophers of neo-Kantian, phenomenological and neo-positivist orientations, who are against a conception of truth as the mirroring of a transcendent reality. This has been stressed again in recent years, only in order to criticize the relativist outcomes of the new philosophy of science. The weight of this kind of difficulty is considerably increased by the following fact: the 'tension' towards objective truth and knowledge, understood as intertheoretical, interlinguistic and intercultural ideals, is demonstrated by the efforts we constantly make to penetrate into other systems of thought, to make them our own through the work of translation and interpretation, and to offer, if necessary, a kind of comparative evaluation of them.

The second difficulty – originally identified by Plato in the *Theaetetus* – consists in the fact that if the supporter of the relativity of truth and objectivity does not consider his position as a mere personal attitude, devoid of any theoretical coercive strength, he ends by putting forward a theory of truth and objectivity which is internally contradictory. 'Truth, says the cultural relativist, is culture-bound. But if it were, then he, within his own culture, ought to see his own culture-bound truth as absolute. He cannot proclaim cultural relativism without rising above it, and he cannot rise above it without giving it up'. These words by Quine (1975, p. 327f) are echoed by those with which Putnam (1983, p. 288) has recalled the answer that an anti-relativist such as Garfinkel is in the position to give the relativist: 'I know where you're coming from, but, you know, Relativism isn't *true-for-me*'.

It seems to me that the internal inconsistency of radical relativism when meant as a theory is one of the reasons why there are not many philosophers willing to proclaim themselves relativists *tout court*. But how can we hope to avoid a radically relativist position, without ignoring all those aspects of cognitive activity that show its dependence on a set of epistemic conditions? This question leads us to one of the three great guiding ideas that, with empiricism and moderate epistemic relativism, constituted the main aspects of the *esprit positif* and that, I think, can be still preserved. This third idea is the idea of a non-metaphysical cognitive objectivity. In my view, the answer to the problem of realism, to the *Realismusfrage* (as Carnap and Reichenbach called it in their letters) depends on the specification of a notion of truth which, although not bound to metaphysical realism and not antithetical to epistemic relativism, will not be relativist to the point of pulverizing the notions of truth and objectivity and turning them into a multiplicity of truths and objectivities, each of which is indisputably such with regard to its own frame of reference.

In my book, I tried to show that it is possible to attain this goal by accommodating the proposal put forward some time ago by Popper precisely in order to contrast radical relativism, within a framework which is different from metaphysical realism. Truth and objectivity must indeed be assumed – as Popper maintained – as unitary ideals or regulative values which transcend the particular presuppositional structures which enframe each concrete cognitive act. Their transcendence, though, is not to be understood in the metaphysical sense of an unknowable correspondence between our cognitive claims and truth in itself. It is possible to see a different way of intending this transcendence, if we place ourselves in a positive, I dare say almost operational, perspective. Since the common and naive notions of truth and objectivity are permeated with realism and correspondentism, it is necessary to partially put aside what we usually mean by truth and objectivity, in order to see which notion of truth and objectivity is inherent in the anti-relativist (or objectivist) tension that informs our cognitive efforts, insofar as these efforts are intentionally aimed at objectivity and knowledge of truth.

It seems to me that this is, on the whole, the direction in which neo-Kantian logical idealism has moved, when it ceased to measure the validity of knowledge according to a standard that transcends it, and it acknowledged – as Cassirer says – that 'science has, and can have, no higher criterion of truth than unity and completeness in the systematic construction of experience as a whole' (Cassirer 1953 [1910], p. 187; translation modified). If we start from the 'positive' idea that knowledge (especially scientific knowledge) is a fact and not a problem, the anti-realist statement that 'we can never compare the *experience* of things with the *things themselves*, as they are assumed to be in themselves separate from all the conditions of experience' (Cassirer 1953 [1910], p. 278), no longer leaves us with radical relativism as the only alternative. We are offered another possibility by theses that have been proposed, although with different goals in view, within phenomenology and logical idealism.

Referring freely to these, it is possible to confer a non-metaphysical value upon the idea of the unitary transcendence of truth and objectivity, interpreting them as regulative ideals which guide scientific and cognitive activity towards the substitution of 'a relatively narrower aspect of experience, by a broader, so that the given data are thereby ordered under a new, *more general point of view*'. Cognitive activity becomes then 'a perpetually self-renewing process with only relative stopping-points' (Cassirer 1953 [1910], p. 278; italics added). It is these very stopping points that give a content, which can be specified only case by case, to the categories of 'objectivity' and 'truth' in their applications to concrete cases and particular judgements. This, though, does not interfere with the unitary transcendence of these concepts as regulative ideals which direct research towards conceptual syntheses which are progressively richer in data, more articulated and more comprehensive.

To put it in Husserl's terms, instead of Cassirer's, objectivity and truth become the ideal unitary correlate of a potentially infinite series of cognitive processes, that intentionally aim at truth and objectivity under the guidance of the regulative ideal of the maximum possible enrichment of the experiences and of the maximum possible integration of the available data and the various perspectives of unification subsequently adopted. What is commonly called the 'real world' becomes 'the ideal limit-plane of resolution of the fragmentary manifold of experience' (Preti 1946, p. 94). From this perspective, also the notions of objectivity and truth, like that of good, become purely formal ideal categories: they become 'empty buckets' which are filled at the river of history. It seems to me that this famous and rather abused metaphor by Simmel expresses very well the tension that must constantly exist in positive philosophy between relativist and historicist instances on the one hand, and objectivist and theoretical-evaluative instances on the other.

One might be tempted to think that my suggestion – which I cannot develop more extensively here – meets with an epistemological obstacle in those very views of history and science which led to radical relativism. In fact, one can easily show that even the conception of objectivity that belongs to Cassirer's logical idealism, no less than that of many versions of convergent realism, relies on the idea of a certain empirical cumulativism and of a certain theoretical continuism. Cassirer himself says that 'the one reality can only be indicated and defined as the ideal limit of the many changing theories ... since only [by the assumption of this limit] is the continuity of experience established' (Cassirer 1953 [1910], p. 321f). Cassirer's logical idealism requires a certain continuity on the theoretical level, a certain constancy of the 'objective' reference. In such a way, his conception extends to the history of science requirements which are superior to those advanced by a phenomenalist like Mach, or by a conventionalist like Poincaré, who limited themselves to requiring a certain continuity on the level of empirical-observational relations. If some scholars have maintained the plurality of truths and worlds, it is because they are persuaded that individuals who speak very different languages, live in different realities, or that 'revolutionary' changes of the disciplinary matrixes entail changes in the universe.

These considerations, in my view, actually hit the weak point of Cassirer's 'liberalization' of the transcendental construal of the epistemological problem. If, in order to keep ourselves faithful to transcendentalism, we maintain the anti-empiricist idea (rightly criticized by Reichenbach) of the 'ultimate logical invariant of experience' (Cassirer 1953 [1910], p. 269; Reichenbach 1978 [1933], p. 401), in logical idealism too, as in convergent metaphysical realism, we require the history of our cognitive efforts to exhibit some form of progressive unveiling of the ultimate structures of reality. It is not very important whether these structures are conceived primarily as properties of the subject, or as properties of the object, or like both these things together: it is always ultimate structures we are dealing with and it is not clear why idealist metaphysics should be better than realist metaphysics. For this reason I do not think we can solve the difficulties raised by the 'new philosophers of science' retaining a neo-Kantian perspective. In any case, I am also convinced that it is exactly the empiricist component of my theoretical reappraisal of the guiding ideas of positive philosophy that saves it from running the same risks run by Cassirer's philosophy. In the context of empiricism and moderate relativism, the interpretation of the unitary character of truth and empirical objectivity as regulative ideals of scientific and cognitive activity loses any metaphysical significance whatsoever and allows us to look at the ironies of history with empiricist detachment.

I think that the problem of incommensurability does not show the impracticability of my point of view, whereas, vice versa, scepticism undermines metaphysical realism at its very roots. The reason for this difference is that my proposal – *qua* proposal with a strong empiricist component – does not advance 'factual' claims which must be transcendentally or metaphysically founded or foundable. My idea does not require that in fact there have not been, there are not, and there cannot be deeply divergent cultures and scientific constructions, inspired by very different ways of looking at the universe. Neither does it require that there be or there not be a theory which constitutes the ideal limit towards which the changing theoretical constructions appearing at the horizon of history tend. In an empiricist's perspective, these occur as empirical questions, and, happily for me, as an empiricist I am not expected to legislate a priori on them. My conception requires only that there be no valid reasons to suppose that it is impossible in principle to establish a comparison between such systems of thought in order to arrive, if we want to, at some comparative judgement, interwoven with empirical and rational motivations (in the sense of an historically conditioned, and thus not absolute rationality – a conception that I cannot specify here).

The important philosophical point that must be stressed is that even the comparability between different conceptual and theoretical systems is something which is not absolute, but relative and dependent on the background assumptions that govern our attempts to translate and to discover adequate hypotheses of connection. As Michael Devitt has noted, it is natural that 'theory comparison must always involve *some* point of view about the domain in question. But this is just to say that theory comparison is theory-laden, which is true even when the most commensurate theories are being compared' (Devitt 1984, p. 154). If we put things in these terms, the problems posited by the incommensurability thesis will no longer be philosophical problems, and so will appear for what they really are, namely logical and scientific problems. Finding solutions for the comparability problems that theories posit in concrete cases will depend upon our capacity to elaborate abstract logical models of intertheoretical relations and develop concrete analyses and scientific hypotheses capable of guaranteeing the individuation of qualified areas of theoretical and empirical intersection which are needed in order to carry out the proper controls. And if, after all, all attempts to build bases of comparability fail, this will only mean that, at least for the time being, one of the two theories talks, in a way which is totally incomprehensible to us, about things whose nature we do not understand. It will not mean that the theory tells us different truths, which cannot be compared with ours, about the things that we do know, nor that the things it tells us about, if they really do exist, cannot be considered as parts of the one reality which constitutes the ideal unitary correlate of our potentially infinite cognitive acts.

Apparently, there does not seem to be anything that can endanger the trans-systemic and intertheoretical regulative ideal of the unitary character of what is objectively valid, when, in an empiricist manner, this ideal is disentangled from metaphysical perspectives concerning the more or less progressive and conjecturally evaluable uncovering of the ultimate structures of the known reality or of the knowing subject. Contrary to what happens in metaphysical realism, where the absence of parameters which could establish the correspondence between knowledge and reality is conducive to scepticism, in a positive perspective it is not necessary to specify in a predictive and general way the instruments which make the various theoretical systems logically and empirically comparable: the unitary character of truth and objectivity, posited as regulative ideals of cognitive activity, are, and must remain, empty buckets. It behoves concrete research to fill these buckets with the contents which are deemed relatively valid in individual cases.

What is required for a philosophical framework of the positive kind is not, after all, the indication of procedures apt to guarantee the comparability between past, present, and future theories, but the proof that this comparability is not impossible in principle. From a positive point of view, characterized by empiricism and moderate relativism, to posit the unitary character of truth and objectivity as regulative ideals means to see as one's goal not the discovery of a privileged language that will be able to say everything about its reference (an idea which is rightly criticized also by Lyotard 1992 [1988], p. 83), but rather the construction of increasingly comprehensive languages, theories, and schemes of connection, which may constitute a relatively neutral basis for the discussion and the evaluation of alternative perspectives. If what follows from the Quinean principle of the systematic indeterminacy of translation is that not even meanings should be considered as something absolute, completely independent of our ways of interpreting them, then also the comparison between particular perspectives and their overcoming within wider perspectives will depend upon the context of research and will have to be construed case by case, according to modalities which are subject to all the rational and empirical constraints which characterize each sector of cognitive investigation.

In this sense, an adequately updated positive perspective seems to be able to encompass both the idea of the unitary character of truth, conceived as a regulative ideal, and the postmodern rejection of a 'unique big narrative' capable of transcending the limitations and the historical conditionings of the cognitive process.

Note

1 Recently Stathis Psillos (1999) provided a powerful and articulate defence of scientific realism against the different forms of scientific anti-realism (instrumentalism, constructive empiricism, Carnap's neutralism, and so on). The problem is that it deals with this question from the point of view of scientific realism: in other words he deals with the relationship between theoretical entities and observable ones, but he does not aim at discussing the question of the general validity of realism and, more particularly, the defence of realism from the attacks of scepticism.

References

Broad, Charlie D. (1978), *Kant. An Introduction*, Cambridge, Cambridge University Press.

Cassirer, Ernst (1922) [1907, 1911], *Das Erkenntnisproblem in der Philosophie und Wissenschaft der neueren Zeit*, 3rd ed., Berlin, Cassirer, vol. 2.

Cassirer, Ernst (1953) [1910], *Substance and Function and Einstein's Theory of Relativity* (1923) [*Substanzbegriff und Funktionsbegriff* (1910) and *Zur Einstein'schen Relativitätstheorie* (1921)], authorized English translation by W. Curtis Swabey and M. Collins Swabey, New York, Dover.

Cassirer, Ernst (1981) [1918], *Kant's Life and Thought* [*Kant's Leben und Lehre*], English translation by J. Haden, Introduction by S. Körner, New Haven, Yale University Press.

Devitt, Michael (1984), *Realism and Truth*, Oxford, Blackwell.

Duhem, Pierre (1962), *The Aim and Structure of Physical Theory* [*La Théorie Physique: Son Objet, Sa Structure* (1904–6, 1914)], English translation by P. P. Wiener, from the 2nd ed. (1914), New York, Atheneum.

Feyerabend, Paul K. (1987), 'Consolations for the Specialist' (1970), in Lakatos and Musgrave (eds), pp. 197–230, 1987.

Fine, Arthur (1984), 'The Natural Ontological Attitude', in *Scientific Realism*, Jarrett Leplin (ed.), pp. 83–107, 1984.

Fleck, Ludwik (1979) [1935], *Genesis and Development of a Scientific Fact* [*Entstehung und Entwicklung einer wissenschaftlichen Tatsache. Einführung in die Lehre vom Denkstil und Denkkollektiv*], T. J. Trenn and R. K. Merton (eds), English translation by F. Bradley and T. J. Trenn, Chicago/London, University of Chicago Press.

Hacking, Ian (1983), *Representing and Intervening. Introductory Topics in the Philosophy of Natural Science*, Cambridge, Cambridge University Press.

Hempel, Carl Gustav (1988), 'Limits of a Deductive Construal of the Function of Scientific Theories', in *Science in Reflection*, 'The Israel Colloquium: Studies in History, Philosophy, and Sociology of Science', Ullmann-Margalit (ed.), vol. 3, Dordrecht, Kluwer, pp. 1–15.

Hempel, Carl Gustav (1992), 'Eino Kaila and Logical Empiricism', in I. Niiniluoto, M. Sintonen, G. H. Von Wright (eds), *Eino Kaila and Logical Empiricism*, 'Acta Philosophica Fennica', 52, pp. 43–51.

Kant, Immanuel (1985) [1781, 1787] [1929, 1933] *Critique of Pure Reason*, [*Kritik der Reinen Vernunft*], English translation by N. Kemp Smith, London, Macmillan.

Kuhn, Thomas S. (1987), 'Reflections on My Critics' (1970), in Lakatos and Musgrave (eds), pp. 231–78, 1987.

Kuhn, Thomas S. (1996) [1962], *The Structure of Scientific Revolutions*, Chicago/ London, University of Chicago Press, 3rd ed.

Lakatos, Imre (1987), 'Falsification and the Methodology of Scientific Research Programmes' (improved version of a paper published in 1968), in Lakatos and Musgrave (eds), pp. 91–196, 1987.

Lakatos, Imre and Musgrave, Alan (eds), (1987) [1970], *Criticism and the Growth of Knowledge. Proceedings of the International Colloquium in the Philosophy of Science*, London 1965, vol. 4, Cambridge, Cambridge University Press.

Laudan, Larry (1984) [1981], 'A Confutation of Convergent Realism', in *Scientific Realism*, Jarrett Leplin (ed.), pp. 218–49.

Leplin, Jarrett (ed.) (1984), *Scientific Realism*, Berkeley/Los Angeles, University of California Press.

Lyotard, Jean-François (1992) [1988], *Peregrinazioni. Legge, forma, evento* [*Peregrinations. Law, Form, Event*], Bologna, Il Mulino.

Papineau, David (1996), 'Philosophy of Science', in *The Blackwell Companion to Philosophy*, N. Bunnin and E. P. Tsui-James (eds), Oxford, Blackwell, pp. 290–324.

Parrini, Paolo (1994a) [1990], 'On Kant's Theory of Knowledge: Truth, Form, Matter' in *Kant and Contemporary Epistemology*, Paolo Parrini (ed.), University of Western Ontario Series in Philosophy of Science, Dordrecht, Kluwer, pp. 195–230.

Parrini, Paolo (1994b), 'With Carnap Beyond Carnap. Metaphysics, Science and the Realism/Instrumentalism Controversy', in W. C. Salmon and G. Wolters (eds) *Logic, Language, and the Structure of Scientific Theories. Proceedings of the Carnap-Reichenbach Centennial, University of Konstanz, 21–4 May 1991*, Pittsburgh/ Konstanz, University of Pittsburgh Press and Universitätsverlag Konstanz, pp. 255–77.

Parrini, Paolo (1998) [1995], *Knowledge and Reality. An Essay in Positive Philosophy* [*Conoscenza e realtà. Saggio di filosofia positiva, Roma-Bari, Laterza*], English translation by Paolo Baracchi, Dordrecht, Kluwer.

Preti, Giulio (1946), 'I limiti del neopositivismo', *Studi Filosofici*, 7, 6, pp. 87–96.

Preti, Giulio (1974), 'Lo scetticismo e il problema della conoscenza', *Rivista [critica] di storia della filosofia*, 29, pp. 3–31; 123–43; 243–63.

Psillos, Stathis (1999), *Scientific Realism: How Science Tracks Truth*, London/New York, Routledge.

Putnam, Hilary (1982), *Reason, Truth and History*, Cambridge, Cambridge University Press.

Putnam, Hilary (1983), *Realism and Reason, Philosophical Papers*, vol. 3, Cambridge, Cambridge University Press.

Quine, Willard V. O. (1975), 'On Empirically Equivalent Systems of the World', *Erkenntnis*, 9, pp. 313–28.

Reichenbach, Hans (1978) [1933], 'Kant and Natural Science' ['Kant und die Naturwissenschaft'], in H. Reichenbach, *Selected Writings*, vol. 1, M. Reichenbach and R. S. Cohen (eds), Dordrecht, Reidel, pp. 389–404.

Rey, Abel (1907), *La Théorie de la physique chez les physiciens contemporains*, Paris, Alcan.

Russell, Bertrand (1990) [1899], 'The Axioms of Geometry' ['Sur les Axiomes de la Géométrie'], in B. Russell, *Philosophical Papers 1896–99*, N. Griffin and A. C. Lewis (eds), textual apparatus prepared by W. G. Stratton, London, Unwin Hyman, pp. 390–415 [French text on pp. 432–51].

Sapir, Edward (1951) 'The Status of Linguistics as a Science' (1929, in E. Sapir, *Selected Writings of Edward Sapir in Language, Culture and Personality* (1949), D. G. Mandelbaum (ed.), Berkeley/Los Angeles/London, University of California Press), pp. 160–66.

Schlick, Moritz (1974) [1918, 1925], *General Theory of Knowledge* [*Allgemeine Erkenntnislehre*] English translation of the 2nd revised ed. by A. E. Blumberg, with an Introduction by A. E. Blumberg and H. Feigl, Wien/New York, Springer-Verlag.

Schlick, Moritz (1979) [1932, 1938], 'Form and Content. An Introduction to Philosophical Thinking' (1938), in M. Schlick, *Philosophical Papers*, vol. 2 (1925–36), H. L. Mulder and B. F. B. Van De Velde-Schlick (eds), 'Vienna Circle Collection', Dordrecht, Reidel (Kluwer), 1979, pp. 285–369.

Suppe, Frederick (1977) [1973], 'The Search for Philosophical Understanding of Scientific Theories' and 'Afterword — 1977', in *The Structure of Scientific Theories*, edited with a Critical Introduction and Afterword by F. Suppe, Urbana/Chicago/London, University of Illinois Press, pp. 1–241, 615–730.

Wolff, Robert P. (1963), *Kant's Theory of Mental Activity. A Commentary on the Transcendental Analytic of the 'Critique of Pure Reason'*, Cambridge, MA, Harvard University Press.

Chapter Eight

Realism vs Nominalism about the Dispositional–Non-Dispositional Distinction

Mark Sainsbury

Hume on matters of fact and powers

Hume's distinction between relations of ideas and matters of fact has often been appreciated primarily for its alleged epistemological aspects: the former but not the latter are supposed to be knowable a priori. I wish to draw attention to a metaphysical aspect of the distinction. Concerning the relation of identity between the sum of the angles of a triangle and 180°, a relation of ideas, Hume says:

> this relation is invariable, as long as our idea remains the same.

I suggest that this involves a supervenience claim: the identity relation supervenes on being a triangle; that is to say, there is no changing the identity relation between the angles so long as the property of their forming a triangle still holds. Some of the evidence for this reading comes from the almost immediately succeeding passage which introduces matters of fact:

> the relations of contiguity and distance betwixt two objects may be
> changed merely by an alteration of their place, without any change
> on the objects themselves or on their ideas. (T 69)

Hume seems to be telling us that contiguity and distance do not supervene on intrinsic properties: objects can change their relations of contiguity and distance without any change in their intrinsic properties. Given that matters of fact are supposed to be just what relations of ideas are not, this rather clear denial of supervenience supports the interpretation of the other quotation as an affirmation of supervenience.

Causation is supposed to be in the same category as contiguity and distance: a matter of fact rather than a relation of ideas. It ought to follow that causal relations fail to supervene on intrinsic properties. Yet according to the first definition of causation, it may seem that they do supervene on intrinsic properties: there is no changing the pattern of causal relations among objects without changing the intrinsic features of the objects, those which sustain their similarity relations. This thought requires qualification (and will receive it shortly). For the moment it suggests an unsuspected tension in Hume's thought about causation (it both is and is not supervenient upon intrinsic properties) which I wish to compare with his view on powers.

He says that powers are not 'compleat': they 'point out', and are not fixed by the qualities or ideas of the object. By contrast:

> Solidity, extension, motion; these qualities are all compleat in themselves, and never point out any other event which may result from them. (E 52)

But powers are supposed to be qualities, in or 'on' the object which possesses them. They must be intrinsic in order to do the causal and explanatory work they are supposed to do. We seem forced to say that powers are intrinsic and non-intrinsic, relational and non-relational; and this seems to commit one to the view that there are no such things as powers.

Parallel reasoning would lead to the denial of causal relations. But the argument for the supervenience of causation, apart from defects like reliance on the unexplained notion of 'intrinsic', fails to distinguish between the supervenience of the total pattern of causal relations, and the supervenience of specific cases of causation. The total pattern of causal relations does supervene, in Hume's theory, upon the distribution of the intrinsic properties whose sharing or not sharing makes objects similar or dissimilar. It does not follow, and for Hume it is not so, that a specific case of the causal relation between particulars A and B supervenes upon the intrinsic features of A and B. Quite the contrary: whether or not A caused B depends upon how things are with other, similar objects. We can take Hume at his word when he classifies each instance of the causal relation as a matter of fact, understood as a non-supervenient relation.

This leaves logical space for the possibility that Hume perceived an inconsistency in the metaphysics of powers which did not touch his metaphysics of causation. I have not found serious support for this suggestion in his texts, little beyond the frail evidence of two paragraphs back. However, the general issue seems worth exploring for its own sake, and it connects with a contemporary discussion. Dispositional properties seem to point outwards, as Hume said, to the effects they tend to produce; their existence somehow involves more than just the possessor of the property. Non-dispositional properties have a chance of being intrinsic. It is tempting to think of the dispositional as dependent upon the non-dispositional. But if the former have a 'pointing out' feature lacked by the latter, it is hard to see how this dependence can work. To recapitulate the putative Humean problem with powers: dispositions must be non-intrinsic, because they 'point out' to their standard effects; but they must also be intrinsic, in order to be properly dependent upon the non-dispositional.

Dispositional nominalism and dispositional realism

One way to cut through many problems about the relation between the dispositional and the non-dispositional is to hold that these terms do not mark out metaphysically distinct kinds of properties, but mark only a distinction among predicates.[1] On this view, which I call 'dispositional nominalism', the expression 'dispositional property' is misleading. Properties cannot be grouped into the dispositional and the non-dispositional. They can be referred to in different ways, for example by the effects they tend to have or by structures which sustain them, or with which they are identical. If we pick out a property by its typical effects we often use a predicate which deserves to be called dispositional; if we pick out a property in other ways, the predicate we use may be called non-dispositional. This perspective promises to relieve the tension we found in Hume: it is not powers which are or are not 'compleat', or which do or do not 'point out'; these are features of our ways of referring to powers. The same property might be referred to in very different ways, just as we might characterize a person by a feature which 'points out' to another thing (sole occupant of cell 197 on such-and-such a date) or a feature which is 'compleat' (having this fingerprint pattern).

Dispositional nominalism contrasts with 'dispositional realism' according to which dispositional and non-dispositional properties differ in nature, in such a way that no property of one kind could be a property of the other. This is the traditional view, going back at least to Locke, who thought that the secondary qualities could be characterized as nothing but powers, whereas primary qualities were powers plus something else. It leads to familiar problems: How are dispositional properties related to non-dispositional ones? How can dispositional properties be causally efficacious? But if they are, must there not be large-scale overdetermination, since causal efficacy ultimately rests with the non-dispositional? How can we avoid thinking of even non-dispositional properties as dispositional, as we come to appreciate their typical effects? In that case, how can there be any non-dispositional properties? And then what could genuinely occupy space? (Cf. Blackburn 1990, p. 258.)

I will take for granted a form of realism about properties, according to which properties exist independently of our thought and talk. (If one is not realist about properties, one will presumably be already committed to dispositional nominalism.) Although what this amounts to stands in need of analysis, I shall take for granted three approximate guides: (1) we cannot count on every predicate introducing a property; (2) things which differ in properties across times or worlds differ genuinely, and do not merely Cambridge-differ; (3) predicates used essentially in stating the fundamental laws introduce properties. A justification for (1) is that it seems that we can invent new predicates, but a realist about properties should not think that properties can be invented. Turning to (2), there are certainly examples of Cambridge differences which one would be disinclined to call differences of property. For example, suppose there is absolute space, then a change in absolute spatial position seems not to be a change in property: it would make no real difference if the whole universe were to shift an inch to the right. To generalize to the claim that no Cambridge-difference is a difference of property would be rash, but we will hold that generalization in place for the moment. A justification for (3) is that a realist about properties will presumably want them to be engaged in serious causal work; presumably this work is systematically described in the fundamental laws, and so predicates essentially used in such description introduce properties.

We need an account of what it is to introduce a property which is neutral in the debate between dispositional nominalist and dispositional realist. We can start by appealing to what makes an application of the relevant predicate true. In typical cases, the truth-maker will contain a property. There is no guarantee that there is a property which enters into every truth-maker for each of the true applications. I shall take it that a predicate introduces a property iff there is a property which features in every truth-maker for every application of the predicate. Suppose, for example, that 'has weight 10 grams' intuitively does not introduce a property. It is natural to suppose that the truth-making property of objects to which this predicate is truly applied varies from application to application (involving one mass on the Moon, and another on Earth). If so, the proposed test delivers that this predicate does not introduce a property. Suppose that 'having mass 10 grams' does introduce a property. It is natural to think that there is a single property shared by every truth-maker for every application. If this is right, our proposed test again delivers the appropriate verdict.

If a predicate, F, introduces a property, then I shall say that the nominalization 'the property of being F' refers to the introduced property. But some caution is needed not to think of 'reference' in this context as a relation whose obtaining or not is an a priori matter, relative to understanding the predicate. Given realism about properties, there will be no close link between property-introduction and semantics. A predicate's introduction of a property is not a condition of its intelligibility, a property is not a predicate's 'semantic value', and it is not a priori whether or not a predicate introduces a property. By the same token, there is no guarantee of an a priori knowledge of whether nominalizations refer.

One way to bring out one aspect of realism about properties is to insist upon a firm distinction between second-level properties and second-order ones (though it would be idle to pretend that people do or even should keep to just this terminology). A first-level property is a property of individuals, and a second-level property is a property of first-level properties. (If dispositional realism is right, then being a dispositional property is a second-level property if it is a property at all.) A first-order property is one expressible without quantifiers, and a second-order property is one expressible only with quantifiers which range over the things to which the property applies (perhaps like the property of having a brother). A realist about properties ought to be happy with the first distinction but unhappy with the second. There is certainly a distinction between predicates which do not contain quantifiers and those which do, but the realist has no business to suppose that there are properties only expressible using quantifiers, for properties are just entities in the world which we can in principle refer to in no end of different ways. If there were a property which we could refer to, in our language, only by using a quantifier, that would reflect some oddness of our current vocabulary and there could be no inference to a special metaphysical feature of the property itself. In this respect, properties should be just like anything else. You or I might be able to refer to Smith only by quantificational means ('whoever is taller than anyone else in the group') but no realist about individuals could suppose that these quantifiers somehow entered into Smith's nature.[2]

Realism about properties is requisite for a dispositional realist, for whom the distinction between dispositional and non-dispositional properties is a distinction among real things; it is optional for the dispositional nominalist. In order to focus the debate on the dispositional–non-dispositional distinction, I shall assume the dispositional nominalist takes up the option of realism about properties. On this realist assumption, this chapter aims to say something on behalf of dispositional nominalism.

One upshot of the agreed realism about properties is that there are fewer properties than predicates, so some predicates fail to introduce properties, so there will be truths of the form 'there is no such property as being such-and-such' (obtained simply by replacing 'such-and-such' by the predicate in question). Although our language allows us to form from any predicate, F, an apparently referring expression of the form 'the property of being F', it will in every case be a substantive question for the realist about properties, and so for all parties to the dispute upon which I focus, whether this expression really refers to a property.

Metaphysically dispositional: law variations

The distinction between dispositional and non-dispositional properties is most often introduced in linguistic terms. For example, a dispositional property is commonly said to be one analysable in terms of a conditional. Making this intelligible requires some heavyweight metaphysics, for 'conditional' starts life as a property of linguistic things, and work must be done to characterize a corresponding ontological feature. The neutral distinction is that between dispositional and non-dispositional predicates: both parties should be able to agree on this. However, for three connected reasons the dispositional realist cannot simply say that a dispositional property is one introduced by a dispositional predicate, and a non-dispositional property is one introduced by a non-dispositional predicate. The first reason is that making the distinction in this way does not do justice to realism about properties: it ought to be an open question, for any predicate, whether it introduces a property. The second reason is that the proposed basis for the ontological distinction does not guarantee its exclusivity: the dispositional nominalist will say that there is nothing to prevent a property being introduced by both kinds of predicate. The third reason is that it leaves the territory entirely open to the dispositional nominalist who will say: 'I understand the distinction among predicates, but tell me how the distinction is mirrored in the nature of reality. Real things certainly do not, for example, contain conditionals, any more than they contain quantifiers'. Even if the dispositional realist were to start by making the linguistic distinction, she must go on to explain the metaphysical distinction.

I begin with an initial dilemma for the dispositional realist. Many dispositional predicates are 'multiply realized': fragile paper is fragile in one way, fragile chairs in another, so 'fragile' is realized in one way in its application to paper and in another in its application to chairs. If a predicate is multiply realized, it is very unlikely to introduce a property. The multiple realization means that different properties make for the truth of different true ascriptions. Though formally consistent with there being a single property which makes for truth for all true ascriptions, this is a rather threadbare possibility. No realist about properties will allow that arbitrary disjunctions of properties are properties, so a single truth-making property would be an overdetermining truth-maker, and the more attractive truth-making theories have a built-in minimality: a truth-maker is a minimal condition for truth. Setting aside the possibility of a single truth-maker, it follows that multiply realized dispositional predicates do not introduce properties (by the standards set in the previous section). That is one horn of a dilemma, which applies to standard predicates like 'fragile'. It is not that they introduce non-dispositional properties, but that they do not introduce properties at all.

The other horn is that the predicate is not multiply realized, so that it is not the case that different properties make for the truth of true ascriptions. Since normal cases of true atomic ascription require a property as truth-maker, this means that there is a single property which is truth-maker for all these cases. Then indeed the predicate introduces a property, but it now becomes hard to see why it should not be identified with a property introduced by a non-dispositional predicate. The pair 'has mass *n* grams' and 'is disposed to impart so much momentum to a stationary body with which it collides at such-and-such velocity' might be examples. Arguably, there is no multiple realization, but it is hard to resist identification.

This dilemma is, as indicated, not a decisive argument. In particular, no irresistible reason has been given to favour identity. However, a more detailed problem can be sketched for the dispositional realist, once he gives us a better idea of what constitutes his metaphysical distinction.

It may be that there is more than one way to explain this, but in this paper I shall consider just two. The first idea is that the extension of a dispositional property can be changed just by a change in laws, whereas this is not so for a non-dispositional property. We imagine the change in law to be instantaneous, and we can call a world in which such a change occurs a 'law variant'. A property is dispositional iff for some law variant it changes its extension at the instant of the change in laws. (Obviously any property may alter its extension, both absolutely and relative to the extension it would have had the laws not changed, after the change of law.)

Typical glasses are fragile in worlds which share the laws of our world, but some of these are law-variants in which the crystalline structure of glass instantaneously becomes very hard to disrupt, so that glasses instantaneously become non-fragile. This suggests that fragility meets the test for being a dispositional property. By contrast, it would seem that things that are square in our world will be square in any law-variant, which suggests that squareness is a non-dispositional property. These rulings accord with what is intuitively wanted from the distinction.

The main problem for dispositional realism is that the application of the law-variant criterion assumes that there is such a property as being fragile, whereas there are two prima facie reasons for doubting this. The first is that the criterion ensures that no dispositional property can feature in a law; the second is that dispositional differences seem to emerge as Cambridge-differences. These consequences, shortly to be argued for, are hard to reconcile with the view that there really are properties which are dispositional by the proposed test.

Let us think of worlds as built out of states-at-a-time, where a (monadic) state is built out of a property and an object. Simplifying by considering only deterministic laws, worlds coincide in the totality of their laws during a stretch of time t_1 to t_2 only if, if they coincide in all their states at t_1 then they coincide in all their later states at every time until t_2. Laws are relations between states-at-a-time: they are the patterns of unfolding of states-at-a-time. Hence a mere change in laws at a time cannot change the states-at-that-time, and since these are constructed out of properties, a change in law at a time cannot change the distribution of properties at that time (though it will affect how later distributions will be, in contrast to how they would have been but for the change). Hence a property which is dispositional by the proposed test cannot feature in a law.

I took featuring in a law as an important mark of being a property. The mark was offered only as a sufficient condition (predicates with essential use in statements of law introduce properties) but the rationale gives some basis for thinking of it as a necessary condition: properties are to do causal work, and so there is some ground for thinking that every property will feature in some law. Even if this is too strong, it would be quite extraordinary if there were a significant range of properties, the dispositional ones, none of which featured in any law; indeed, which were precluded a priori from featuring in a law. The proposed account of what it is to be a dispositional property provides reasons for thinking that there are no such properties, contrary to the intentions of the dispositional realist.

A second consideration points in the same direction: law-variants arguably exhibit mere Cambridge-difference. The difference in laws ensures that some things will behave differently, and so will in future acquire different properties. However, the difference in laws itself does not seem to touch the nature of the objects: a pair of objects differing only in that one is in a world in which it figures in a certain pattern of law, another in a world in which it does not, resemble a pair of objects differing only in that one figures as the apex of a triangular arrangement of objects and the other does not. If there is anything in the thought that objects differ in properties only if they exhibit 'real' difference and not just Cambridge-difference, it seems to apply here, delivering the result that there are no dispositional properties. (There are, of course, truths and differences involving dispositional predicates, but that does not make for truths and differences involving properties.)

These considerations amount at most to a preliminary case against dispositional realism. Even if there are accepted, there are too many moving parts (the natures of worlds and of laws, for example) to make anything like a refutation of dispositional realism. However, I do think they constitute a challenge, which is reinforced by problems besetting another attempt at making the metaphysical distinction, considered in the next section, and by the happy consequences of dispositional nominalism, some of which are mentioned in the subsequent section.

Dispositional and relational properties

The dispositional realist may suggest that what has gone wrong is that an inappropriate definition of dispositional property has been applied. He might prefer to exploit the 'pointing out' nature of dispositional properties, their relationality. Let us say that a relation is something that takes two objects (its 'terms') to form a state. We now need to think of states as potentially extended in time, since for all we know a relation may hold between things at different times. A relational property is a relation together with just one term, the 'fixed' term. ('Together' here must be understood to indicate that relation and fixed term are related in a relational property as they are in a relational state: predicative cement is required.) The suggestion to be considered in this section is that it is a necessary condition for a property to be dispositional that it be relational, but that typical non-dispositional properties are non-relational.

The fixed term in the relational property will presumably be or be intimately connected with the disposition's standard effect. This cannot be thought of as an individual, since a dispositional realist will want things with different individual effects (though the same kind of effect) to share dispositional properties. The fixed term could perhaps be thought of as itself a property, the property distinctively manifested by the bearer of a dispositional property when the disposition is realized or made manifest. So the supposed dispositional property of being fragile will be a relational property whose fixed term is (something like) being broken. By contrast, a paradigm of a supposedly non-dispositional property, like being square, is not a relational property at all, and has no fixed term.

As before, the problem is whether on this account there really are any dispositional properties. It is not obvious that a proper metaphysics should admit relational properties. I have assumed that the metaphysics will emerge from an account of truth-making. On the view I shall assume, relational states are needed to make true some fundamental relational sentences, and these states require the relation and its terms. None of this requires any commitment to further entities, relational properties. They do not emerge automatically from the existence of the relation and the existence of the fixed term, for to form a relational property the fixed term needs to be cemented to the relation in the way that it is in a state. Moreover, to allow relational properties would be to allow overdetermination among truth-makers. Some atom of the form '*Rab*' would be made true not only by the relational state of *a*-having-*R*-to-*B*, but by the property *a* has of being-*R*-to-*b*, and by the property *b* has of having-*a*-*R*-to-it. So one danger is that the account simply fails to deliver any dispositional properties through failing to make room for relational properties.

But suppose this danger is illusory, for example because it turns out that there are many relational states and wherever there is a binary relational state there are also the two relational properties. The account faces difficulties from the fixed term. The suggestion was that this should be a property which is distinctive of the manifestation of the disposition, but now that we are in the world of serious realism about properties, this is no minor assumption. Have we any reason to think that there is a property shared by manifestations of fragility in bridges, paper, and glass, but not shared by just anything that is failing to function as it should? These things of course 'have something in common', being fractured or torn or broken, but this is not sufficient for the existence of a property, else there would be as many properties as predicates. It is also true that things alike as possible save that one is broken and the other is not differ genuinely, and not just Cambridge-style, so being broken is not excluded from being a property by the Cambridge test. However, it is very unlikely to count as a property in virtue of featuring in fundamental laws. This is not to say that it cannot be a property: that could be decided at best in the presence of a full metaphysics of properties (and this will still not decide the issue, for non-a priori facts will also be relevant). But it is worth stressing the fragile thread upon which dispositional realism would hang were it to go down this road.

It would be foolish to suppose that this consideration is decisive. It is offered simply as a challenge to the dispositional realist to provide a better account.

How dispositional nominalism can solve problems

(1) It is not so very easy to say what dispositional predicates are. Familiar suggestions are that they are ones which are, or are equivalent to, conditionals specifying what does or would happen under certain conditions; or that they are, or are equivalent to, ones of the form 'is disposed to behave thus under such-and-such conditions'. The problems are familiar: (a) not all fragile things (to take one example) are fragile in the same way, so it is hard to specify appropriate conditions (ones just right for manifesting the fragility of glasses, paper and bridges); (b) there are angelic cases, in which angels intervene to prevent typical manifestations; (c) there are finkish[3] cases in which the conditions, in addition to meeting the target specification, contain the power to alter the object's dispositions; and no doubt there are other problems.

It may be possible adequately to resolve these, but the problem will be perceived differently by the dispositional realist and the dispositional nominalist. Suppose the dispositional realist uses being introducible by a dispositional predicate in the account of a property's dispositionality. Then it will matter greatly to get the account of a dispositional predicate right. For the dispositional nominalist, on the other hand, the distinction may be of little importance. At most it would be an issue in semantics, and not in metaphysics.

(2) Did the glass break because it was fragile or because it had some underlying constitution in virtue of which it was fragile? Or was the breaking overdetermined? These are real questions for the dispositional realist. Either there is no such property as fragility, which is in tension with it apparently having a causal role; or there is such a property which must be distinct from all non-dispositional properties, and this raises the question of how it can be causally efficacious without being causally overdetermining. The dispositional nominalist has an option for answering these questions that is not available to the dispositional realist. It arises on

the supposition that 'fragile' does introduce a property, so one can expect it to be causally efficacious. The dispositional realist must say how the property of fragility relates to the non-dispositional properties, for it is natural to think that the non-dispositional ones are jointly causally sufficient for everything. This problem evidently does not arise in this form for the dispositional nominalist. The property of fragility is neither dispositional nor non-dispositional; it can be identical to some property which would properly be specified in non-dispositional terms. If fragility is not a property, the dispositional realist and the dispositional nominalist have a similar problem in accounting for its apparent causal efficacy. In general, a predicate which figures in a causal truth but which does not introduce a property must presumably have, on that occasion, a truth-maker containing a causally efficacious property. There are various ways of developing this story, and many of these variants are available to both parties to our dispute.

(3) Having a specific mass, say *n* grams, is associated with a range of dispositional predicates each of the (roughly indicated) form: 'is disposed to cause such-and-such change of a velocity in a body with which it collides (as a function of the mass of the other body and the velocities of both)'. There is such a predicate (in fact more than one) for each expression for a number. Dispositional nominalists and realists can agree that they do not have to suppose that any or all of these predicates introduce properties. But suppose they do think that some or all of them do introduce properties. The dispositional nominalist can say that they all introduce the same property, that of having mass *n* grams. The dispositional realist is unlikely to be able to say this, for she is likely to think that having mass *n* grams is non-dispositional whereas the property of being disposed to cause such-and-such change of velocity is dispositional; and since she thinks that being dispositional and being non-dispositional are exclusive characteristics, this means that the properties cannot be identical. Yet if they are distinct, there is competition for causal efficacy: which is really doing the causal work in a collision? Or do both the mass and the disposition? In the latter case, is the causation overdetermined?

(4) Prior, Pargetter, and Jackson (1982) offered a reason for thinking that a disposition cannot be identified with its basis. In the terminology of this chapter, with explicit neutrality on the question of dispositional nominalism versus dispositional realism, the claim is that no sentence of the following form is true, if '*F*' is a dispositional predicate and '*B*' a predicate of 'causal basis':

the property of being *F* is identical to the property of being *G*.

The reason for the claim is simply that a disposition may have a different basis in different objects: it is one thing for a chair to be fragile, another for a sheet of paper to be. Prior et al. take for granted that every predicate rightly defined in terms of a subjunctive conditional refers to a dispositional property (p. 254). If an identity sentence like the one just mentioned fails to be true, their explanation is that there are different properties at issue. An alternative explanation is that there is no property of being *F*. The explanation is available to both the dispositional realist and the dispositional nominalist, as I have set up these positions, for neither is willing incautiously to infer to the nature of reality from the nature of language. The dispositional realist, however, has a quite general reason for thinking that no such identities could be true: being dispositional and being non-dispositional are exclusive characteristics of properties. The dispositional nominalist, by contrast, has no general grounds for dismissing such identities. Indeed, it would be odd if there were no truths of property identity, for his claim is that one can refer to one and the same property in various ways. One would accordingly expect there to be examples, and Prior et al. at least rule out a range of putative candidates.

However, it is not hard to find others. We have seen some examples in the case of mass and associated dispositional predicates. More generally, many properties, like being a gene or being an electron, are often in the first instance identified by the causal work they are disposed to do. Often a less dispositional specification becomes available with theoretical advance. Told in one form or another, this is a familiar story, but one whose telling by the dispositional nominalist seems especially straightforward. The theoretical advance gives us new ways of thinking about a previously ill-understood property; it does not discover new properties.

Conclusion

Hume may have felt that powers were under incompatible obligations: to 'point out' and to be 'compleat'. This tension persists for dispositional realists. But a solution is to hand in the shape of dispositional nominalism. I have not done anything which could count as establishing this position here; but I hope I have said enough both to clarify it and to make it an attractive option.[4]

Notes

1 The view goes back at least to Mackie (1973): 'it may well be that most properties ... are ... unavoidably described and introduced in the dispositional style. But this would not make *what is there* dispositional, and it would not give the concept of disposition or power any ontological or metaphysical role' (p. 136). Something like dispositional nominalism has been affirmed by several others, for example Mumford (1998). His version has the disadvantage (from my point of view) of being defended in terms of identities of 'property instances' rather than properties (cf. pp. 145, 159–62).

2 The example recalls Ramsey's discussion of the ontology of Russell's ramified theory of types (1925, p. 192; cf. also 'elementary is not really a characteristic of the proposition, but of its mode of expression', p. 186). Ramsey in effect said that if the types were real non-linguistic things, there could be no justifiable constraint on what means, quantificational or other, were used to speak of them.

3 In philosophy, a finkish disposition is one which would vanish if put to the test. For example, this glass might be fragile, yet some powerful creature might ensure that dropping it makes its fragility temporarily disappear. In dated slang, a 'fink' is a person who continues to work when his fellow workers are on strike. One can dimly see the connection: a finkish disposition continues to be present even though it doesn't 'work'.

4 A version of this chapter was given as a paper at the conference 'The Problem of Realism' at Genoa, 29–30 November 1999. In the following month I heard a paper given at King's College London by Hugh Mellor, a descendant of which has now been published (Mellor 2000). Mellor also favours dispositional nominalism.

References

Blackburn, Simon (1990), 'Filling in Space'. Reprited from Analysis in S. Blackburn, *Essays in Quasi-Realism*, Oxford, Oxford University Press, 1993, pp. 255–8.

Mackie, John (1973), *Truth, Probability and Paradox*, Oxford, Clarendon Press.

Mellor, D. H. (2000), 'The Semantics and Ontology of Dispositions', *Mind*, 436, pp. 757–80.

Mumford, Stephen (1998), *Dispositions*, Oxford, Clarendon Press.

Prior, E. W., Pargetter, R. J. and Jackson, F. C. (1982), 'Three Theses about Dispositions', *American Philosophical Quarterly* 19, pp. 251–7.

Ramsey, Frank (1925), 'The Foundations of Mathematics', in *Foundations: Essays in Philosophy, Logic, Mathematics and Economics*, D. H. Mellor (ed.), London, Routledge and Kegan Paul, 1978.

Chapter Nine

Scientific Realism without Reference

Pierre Cruse and David Papineau

Scientific realism, normally construed, is a multi-component thesis about the metaphysics, semantics, and epistemology of scientific theories. Roughly speaking, a realist is one who believes that:

(1) there is a mind-independent world for theories to be about, partly containing things not directly available to unaided observation;

(2) scientific theories are able to represent, correctly or incorrectly, facts about this domain; and

(3) we have reason to believe that our most successful theories are at least approximately true descriptions of these facts.

In this chapter the first two claims will not be at issue, and we will assume that they are true. The question under discussion will be how one might argue for the third, epistemological, claim.

Our aim will be to show that a commonly held view on this issue is incorrect. According to this view, it is crucial to the defence of epistemological scientific realism that particular conclusions about the referential status of theoretical terms be justifiable. In order to show that this view is mistaken, we will suggest a version of the realist thesis to which the referential status of theoretical terms is irrelevant. On our version of scientific realism, the cognitive content of a scientific theory lies in its Ramsey-sentence.

We will then consider whether our alternative realist thesis is equivalent to scientific realism more standardly construed. This raises some complex issues about reference. In the end, we argue that our Ramsey-sentence realism does come to the same thing as standard realism. But this further equivalence claim is independent of our main thesis, which is that Ramsey-sentence realism is a substantial version of realism in its own right, and describes the greatest cognitive achievement we are entitled to expect of our theories.

Does approximate truth require reference?

Let us first see why the referential status of theoretical terms is generally thought to matter to scientific realism. A standard line of thought is that the notion of reference enters the fray through abductive arguments which purport to justify realism. Such arguments work by construing realism as a quasi-empirical hypothesis, justifiable on the grounds that it best explains some independently manifest feature of science. The simplest and most well-known argument of this sort (and the one we will concentrate on in this chapter) is the so-called 'no miracles' argument. According to this argument the 'observable' fact that certain theories are empirically successful justifies the hypothesis that they are substantially accurate in respect of their unobservable or deep-structural claims, on the grounds that, if this were not the case, their empirical success would be miraculous. Less contentiously, perhaps, one can construe the argument as the claim that the best explanation for the success of successful theories is that they are approximately true. Although one could justifiably question this sort of argument, we will not do so in this chapter. Our interest is rather in the way it is generally thought to bring in questions of reference.

Clearly, if you want to argue that 'approximate truth' best explains the success of science, the viability of your argument will have to depend on your notion of approximate truth. It is at this point that referential issues are normally brought in. Consider Putnam:

> If there are such things [as electrons, DNA molecules, curved space-time and so on], then a natural explanation of the success of these theories is that they are partially true accounts of how they behave ... But if these objects don't really exist at all, then it is a miracle that a theory which speaks of gravitational action at a distance successfully predicts phenomena; it is a miracle that a theory which speaks of curved space-time successfully predicts phenomena ... (1978, p. 19)

Putnam is saying that successful theories must achieve two things if their success is not to be a miracle. First, the theories must succeed in the identification or postulation of existing theoretical entities, such as DNA or curved space-time. Second, the theories must give an approximately correct description of these entities. The success of a theory that has failed to achieve either of these would, according to this line of thought, be a miracle. Thus, holds Putnam, we are entitled to assume that successful theories generally achieve both.

Reference gets into the act on the back of the first of these requirements: presumably the entities that a theory 'identifies', if it identifies any at all, are the things that its terms refer to. Thus if a theory's terms fail to refer, it fails to identify existing entities, and, so the argument goes, its success would be a miracle. Thus reference appears to be a necessary condition for explaining empirical success via the approximate truth of theories.

Exactly the same point has also been used to argue against realism, the canonical case being Larry Laudan's 'Confutation of Convergent Realism' (Laudan 1981). Laudan concurs with the thought that a realist explanation of empirical success in terms of approximate truth requires that the relevant theory refers, claiming that

> a realist would never want to say that a theory is approximately true
> if its central terms fail to refer.

But he holds that it is simply a datum that central terms in many successful theories do not refer: 'aether', 'phlogiston', 'caloric', and various others fall into this category on his view. Laudan claims that this evidence of 'reference failure' undermines the ascription of approximate truth to theories. If reference failure is widespread, then by the realist's own lights approximate truth cannot be the correct explanation of empirical success.

The state of play between Laudan and the realist therefore appears to be a stand-off; a case of one person's *modus ponens* being another's *modus tollens*. Both parties agree that successful reference is an essential part of what it would take for a theory to be deemed approximately true in a manner satisfactory for a realist explanation of the success of science. The realist claims that the success of successful theories warrants us in claiming that they are approximately true, and thus, that they successfully refer. Laudan, on the other hand, takes reference failure to be an independently established fact, arguing on these grounds that the theories which it afflicts cannot be approximately true, and thus, that success on its own cannot warrant the attribution of approximate truth to those theories.

Can generous theories of reference save realism?

Laudan's case against realism rests on his list of putative reference failures in the history of empirically successful scientific theories: 'aether', 'phlogiston', 'caloric', and so on. In response to (or perhaps in anticipation of) such cases, realists have favoured 'generous' theories of reference which are designed to keep such reference failures to a minimum; theories of reference, that is, on which referential success is the overwhelming norm. Canonical in this regard is the so-called 'causal theory of reference' of Kripke and Putnam, on which theoretical terms are seen as acquiring referents through 'baptisms' or 'dubbings', in which referents are identified via ostention, in combination with a small amount of minimally specific descriptive material; later uses of the term refer to the same thing when an appropriate sort of causal connection obtains with the original baptism. By divorcing reference as far as possible from the theoretical beliefs of speakers who use terms, the theory promises to yield the sorts of consequences the realist would want. There are a number of variations on the causal theory in the literature, and also other generous theories of reference designed to defend the realist picture in a similar way (for example Newton-Smith 1981, Hardin and Rosenberg 1982, Kitcher 1993, and others).

It seems unlikely, however, that any good theory of reference will be able to deal with all Laudan's cases. Take the nineteenth-century 'aether' as an example. As a number of writers have noted (for example Laudan 1981, Hardin and Rosenberg 1982, Worrall 1989), this is a very hard case for the realist. The nineteenth-century theory that explained electromagnetic phenomena as waves in a continuous elastic solid had a welter of notable empirical successes. Yet the received post-Einsteinian view is undoubtedly that no such aether exists.

True, Hardin and Rosenberg (1982) argue precisely that 'aether' was genuinely referential, on the basis of a theory of reference which takes terms to refer to that entity, property, or substance which plays a certain 'causal role'. The aether is thus identified as the bearer of optical phenomena. Unfortunately, Hardin's and Rosenberg's view of reference is far too generous to be plausible. Aside from their non-standard view of the referent of 'aether', they have to wrestle with the fact, pointed out by Laudan (1984), that systematic application of the required principle generates even worse cases: if 'aether' refers to the electromagnetic field on the grounds that it identifies, but misdescribes, the seat of optical phenomena, why does Aristotle's conception of 'natural place' not refer to gravitational attraction on the grounds that it identifies, but misdescribes, the 'cause of fall'? A theory of reference with such consequences would seem to violate all the intuitions that are standardly taken to motivate theories of reference in the first place.

Ramsey-sentence realism

As we stated at the outset, our intention is to propose an alternative realist hypothesis which removes theories of reference from their alleged role in the realist's explanatory scheme. This alternative hypothesis is that the empirical success of scientific theories can adequately be explained by appeal to the approximate truth of their Ramsey-sentences. It is a further question whether this Ramsey-sentence realism is the same as standard scientific-theory realism. This depends on whether the approximate of a Ramsey-sentence is the same thing as the approximate truth of the original theory, and we shall discuss this issue later. Our first task, though, is to explain Ramsey-sentence realism itself.

The procedure for forming Ramsey-sentences is well known. In order to form the Ramsey-sentence of a theory, we replace each distinct theoretical term-type occurring in the theory with a distinct variable. We then place a distinct existential quantifier at the front of the resulting expression for each distinct variable.

Thus if the original theory could be expressed,

$$T(F_1, \ldots, F_n),$$

its Ramsey-sentence would be

$$(E!x_1) \ldots (E!x_n) \, T(x_1, \ldots, x_n).$$

Whereas the original theory would have said, of the putative referents of its theoretical terms, that those things had certain properties, the Ramsey-sentence just says that there exist unique things with those properties.

The first thing to note is that the referential success or failure of the theoretical terms in a theory is irrelevant to the approximate truth of its Ramsey-sentence, since those terms do not occur in the Ramsey-sentence. For a scientific theory itself to be approximately true, the referents of its terms must approximately satisfy the properties the theory attributes to them. If there are no such referents, then the theory cannot be approximately true. But for the theory's Ramsey-sentence to be approximately true, we need only require the approximate truth of the existential claim, that there exist things with such-and-such properties, and this claim could be approximately true even if the theory's terms fail of reference.

For example, suppose 'the aether' fails to refer to anything. Then there is no question of 'the aether' possessing any properties, approximately or otherwise. But this does not preclude the assessment of the relevant existential Ramsey-sentence for approximate truth. It could be approximately true that there is an entity which is the seat of electromagnetic phenomena, and involves transverse radiation, and consists of an elastic solid. After all, there is indeed something which is the seat of electromagnetic phenomena, and involves transverse radiation, and so on – namely, the electromagnetic field – even if it is not an elastic solid. Many of the implications of a Ramsey-sentence could thus still be true, even if the original theory's terms fail of reference. We take this to illustrate the way in which a Ramsey-sentence can be approximately true even if its original theory suffers reference failure.

This possibility now offers an alternative way to argue from empirical success to scientific realism. Faced with an empirically successful theory, the realist can argue, not to the approximate truth of the theory itself, but to the approximate truth of its Ramsey-sentence. Maybe Laudan can discredit the approximate truth of theories as the best explanation of empirical success, on the grounds that many empirically successful theories failed to refer and so cannot be approximately true. But this does not discredit the approximate truth of their Ramsey-sentences as the best explanation of their success. And such explanations do seem highly plausible. Surely the reason nineteenth-century electromagnetic theory worked so well is that its existential claims were approximately true. It would have been a miracle for it to make so many successful predictions unless at least this much were true.

Thus Ramsey-sentence realism. The Ramsey-sentence realist says that we should believe in the approximate truth of a successful theory's Ramsey-sentences, on the grounds that it would be a miracle that the theory were successful, were its Ramsey-sentence not true.

On this conception of scientific realism, debates between 'generous' and 'stingy' theories of reference turn out to be irrelevant. Suppose we are forced to adopt a stingy account of reference on which successful theories characteristically fail to refer. This will not count against a Ramsey-sentence realist explanation of empirically success. For, even given a stingy account of reference, we can still uphold the approximate truth of those successful theories' Ramsey-sentences.

It is worth being clear exactly what Ramsey-sentence realism achieves in the face of Laudan's 'confutation'. In particular, note that it does not immunize realism against any historical evidence whatsoever. Imagine a neo-Laudanian argument which adduced a list of examples showing that the Ramsey-sentences of many past successful theories were not in fact approximately true. If this was historically accurate, then it would amount to a successful confutation of Ramsey-sentence realism. It would discredit the inference from a theory's empirical success to the approximate truth of its Ramsey-sentence, by showing that there are many cases where this inference fails.

But this is no objection to what we have argued. Our view that the realist content of a theory lies in its Ramsey-sentence is not supposed to provide a sure-fire method of rebutting any historical evidence. Ramsey-sentence realism is still hostage to the historical claim that there is approximate truth at the level of Ramsey-sentences, at least for empirically successful theories. As it happens, we do think that the history of science displays such approximate truth, but this would have to be defended elsewhere. Our current intention is only to deliver realism from a spurious argument, namely, that it is incompatible with reference failure.

Are Ramsey-sentences really realist?

Some readers familiar with Ramsey-sentences may be suspicious of their realist credentials. Exactly what a Ramsey-sentence asserts is sensitive to what counts as a 'theoretical' term, since this decides what is to be eliminated from the theory in forming its Ramsey-sentence. The trouble is that some ways of identifying 'theoretical terms' threaten to remove any realist content from Ramsey-sentences.

The most familiar way of drawing the distinction between 'theoretical' and other terms is that proposed by the logical empiricists. The logical empiricists viewed Ramsey-sentences as a way of expressing the empirical content of theories using only 'observational' vocabulary that referred to items or properties that are directly perceivable; thus, all but the observational and purely logical vocabulary in a theory was to be eliminated in forming its Ramsey-sentence.

However, this conception of Ramsey-sentences is unlikely to be helpful to someone who wants to use Ramsey-sentences in a defence of realism. Without going into detail, a now well-known argument shows that Ramsey-sentences as conceived by the logical empiricists make no claims about unobservables. So the claim that such Ramsey-sentences are approximately true can scarcely amount to a version of scientific realism. A detailed discussion of the relevant argument can be found in Demopoulos and Friedman (1985); we will mention it no further here.

Still, there is no need for the realist to be tied to the particular logical empiricist understanding of 'theoretical' as 'non-observational'. Instead, realists ought to invoke the distinction proposed by David Lewis in his article 'How to Define Theoretical Terms'. Lewis suggests that the 'theoretical terms' of a theory ought not to be understood as contrasting with 'observation' terms, but with 'antecedently understood' terms. The theoretical terms (or T-terms) to be eliminated in forming the Ramsey-sentence of a theory are just those whose meaning derives from their place in the theory, while the antecedently understood terms (O-terms, or 'old' terms) are those whose meaning is fixed independently of their place in the theory. Thus the process of Ramsification does not explain how theoretical terms depend for their meaning on observational terms; it only explains how terms newly introduced in a theory depend for their meaning on terms understood independently of that theory.

Lewis puts his distinction forward as a theory-relative one, and claims to commit himself to no absolute distinction between theoretical and non-theoretical terms. However, it is easy to see that this diagnosis of the situation cannot be quite correct. The theoretical terms in any theory T will on Lewis's account have their meanings fixed by their place in that theory and by the meaning of the antecedently understood vocabulary. But that antecedently understood vocabulary might itself depend in meaning on its place in a further theory T', and on T''s antecedently understood vocabulary. Moreover, by conjoining the original theory T with the theory T', and eliminating all of either theory's vocabulary in favour of existentially quantified variables, the original theory T can have its content expressed in O' vocabulary: vocabulary defined in neither T nor T'.

By tracing the relations of definitional dependence backwards in this way, one of three things will happen:

(1) we get into an infinite regress, each term defined in a further theory;

(2) we go round in circles, and every term turns out to be theoretical in the theory in which it occurs; or

(3) we arrive at a set of vocabulary in terms of which any theory can be expressed in Ramsey-sentence form, but which is not defined in any theory.

(1) seems absurd if we can only possess a finite number of theories; and (2) appears to lead to the conclusion that the content of any theory is expressed by a sentence which eliminates all non-logical vocabulary, apparently rendering every theory trivially true (although Lewis, 1984, suggests that this might be his own view). Thus a realist would be well advised to accept the third option, on which theories can have their contents specified using a set of vocabulary which is absolutely non-theory-dependent in meaning.

This might seem effectively to return us to the unhelpful empiricist equation of 'theoretical' with non-observational. This appearance, however, would be deceptive. The only requirement now being imposed on non-theoretical terms is that they are not understood as denoting just those things which satisfy some particular theory. It would require extra empiricist presuppositions to infer from this that the relevant vocabulary must be 'observational' in any substantial sense. Without prior empiricist prejudices, why not allow that a term could fail to be defined in a theory, and yet be neither observational nor logical? Antecedently understood terms could thus refer to such substantial non-logical relations as causation or correlation, or indeed to many kinds of unobservable things.

If this is right, then we can evade Demopoulos's and Friedman's argument that Ramsey-sentences cannot make claims about the unobservable. This means that the approximate truth of a Ramsey-sentence can be a substantial realist claim. The Ramsey-sentence of a theory essentially defines its unobservable descriptive content. It specifies in qualitative existential terms how the theory takes the unobservable world to be. So the approximate truth of this unobservable content can yield a genuinely realist explanation for the empirical success of the relevant theory.

Are Ramsey-sentences different from their theories?

So far we have defended a version of scientific realism by driving a wedge between scientific theories and their Ramsey-sentences. Scientific theories use theoretical terms, and cease to be candidates for approximate truth if those terms fail to refer. Ramsey-sentences existentially quantify those terms away, and so their approximate truth does not depend on questions of reference.

In these last two sections we shall consider whether this wedge between scientific theories and their Ramsey-sentences can be removed. That is, we shall consider whether Ramsey-sentences coincide in content with their theoretical originals. If they do, then our Ramsey-sentence realism will be no different from scientific realism as standardly conceived after all. Our argument for the approximate truth of Ramsey-sentences will simultaneously be an argument for the approximate truth of scientific theories.

The point of focusing on Ramsey-sentences has been to free scientific realism from dependency on disputable theories of reference. Since Ramsey-sentences do not use theoretical terms, Ramsey-sentence realism does not require any claims about the reference of terms. By the same coin, any argument that scientific theories are equivalent to their Ramsey-sentences will have to involve itself with questions of reference. We will need to show that the referential values of the reintroduced terms are such as to render the term-using theories equivalent to their Ramsey-sentences.

So the arguments in these last two sections will hinge on views about the referential workings of theoretical terms in science. Let us emphasize once more that the rest of this chapter is independent of any such views about reference. We need to engage with questions of reference in order to argue that Ramsey-sentence realism is the same as standard scientific realism. But Ramsey-sentence realism itself is independent of any questions of reference.

Having focused the issue, we would like to start by observing that it would intuitively seem rather odd if scientific theories said something different from their Ramsey-sentences. We have argued that the best explanation of empirical success in science is that scientists are characteristically approximately right when they suppose that there exist unobservable entities which bear such-and-such relations to each other and to antecedently available entities. If scientists are nevertheless wrong in their theoretical beliefs, this can only be because their theories are saying something different – presumably that certain other putative entities, so to speak, bear those relations to each other and to antecedently available entities. But why should scientists so want to say something different? Surely the whole idea of a scientific theory about unobservables is to posit hidden causes of a certain kind. But that is just what a Ramsey-sentence does. It is thus difficult to see what point there would be to having scientific theories which say something different from their Ramsey-sentences.

Still, these rhetorical remarks scarcely decide the issue. To clarify the point, note that it is perfectly possible for a term-involving claim to differ in content from its Ramsey-sentence. Suppose we are thinking about the children in a school, and adopt the view 'John is tall, thin, and red-haired'. The 'Ramsey-sentence' of this claim would be 'There exists a unique boy who is tall, thin, and red-haired'. Now, this 'Ramsey-sentence' could be true – there could be a unique tall, thin, red-haired boy in the school – yet my original belief wrong, simply because 'John' names some other boy. Clearly here the original theory – 'John is tall, thin, and red-haired' – differs in content from its 'Ramsey-sentence' – 'There exists a unique boy who is tall, thin, and red-haired'.

This example shows how a claim can differ from its Ramsey-sentence. At the same time, it indicates a model of reference which would allow scientific theories to be equivalent to their Ramsey-sentences. If the terms in a scientific theory are definitionally equivalent to descriptions, and these descriptions derive from the assumptions in the theory, then the theory and its Ramsey-sentence will be equivalent. Thus, if 'John' is equivalent to 'the unique tall, thin, red-haired boy', then the original theoretical claim 'John is tall, thin, and red-haired' collapses into its Ramsey-sentence 'There exists a unique boy who is tall, thin, and red-haired'. A similar descriptive account of theoretical terms would render scientific theories equivalent to their Ramsey-sentences.

How plausible is it that scientific terms are defined descriptively in this way? If the alternative is a causal theory of reference along Kripke–Putnam lines, there are two immediate points that can be made in favour of the descriptive account.

First, the Kripke–Putnam causal account has trouble with terms that are introduced before there is any direct experimental manifestation of their referents. On the Kripke–Putnam view, the referent of a term is fixed via a 'baptism' or 'dubbing' involving some manifestation of the referent. Thus, for example, 'electricity' might be introduced as 'the cause of *those* effects'. But this story clearly will not work with terms like 'positron', 'neutrino', and 'quark', since these terms were explicitly introduced to refer to hypothetical entities that play theoretically specified roles, before any direct experimental manifestations were available for any dubbing ceremony. Cases like these seem tailor-made for a descriptivist account, on which scientific terms denote precisely those entities that satisfy some set of theoretical descriptions.

Second, there is the point, mentioned earlier, that causal theories seem overly generous. If we take the Kripke–Putnam approach to reference seriously, then we seem in danger of having 'natural place' refer to gravitational attraction, and 'phlogiston' refer to absence of oxygen, and so on. Intuitively, it seems wrong to allow that theoretical terms always refer to the causes of the phenomena they were introduced to explain, however mistaken the surrounding theories may have been about the nature of those causes.

Still, even if these two points give us some initial reason to favour descriptivist accounts of scientific terms over causal accounts, we are still some way short of the thesis that scientific theories are equivalent to their Ramsey-sentences.

One problem is that the specific descriptivist account have offered so far also has highly counter-intuitive consequences. Where Kripke–Putnam causal theories are clearly too generous, this descriptivist account is clearly too stingy. Our suggestion was that the whole scientific theory involving a term should define it. The term thus denotes that unique entity, if any, which satisfies all the assumptions the theory makes about it. But, if we take this seriously, then no scientific term will ever denote anything. For no extant scientific theory has ever been right in all its assumptions about the entities it postulates.

The obvious remedy, if we want to pursue descriptivist accounts of scientific terms, is to suppose that only some of the assumptions in scientific theories contribute to the definitions of their theoretical terms. This then promises to allow that some scientific terms ('natural place', 'phlogiston') did indeed fail to refer, without immediately flopping over to the excessively stingy conclusion that no scientific terms ever refer.

But what principles decide which assumptions in a theory contribute to the definitions of its terms, and which do not? The idea of such a division seems dangerously close to an analytic–synthetic distinction. This raises a number of large issues, about which one of us has written elsewhere (Papineau 1996). Without wanting to go into details, the conclusion there reached is that it is a vague matter, within limits, exactly which theoretical assumptions contribute to the definitions of scientific terms.

If this is right, it is then also a vague matter exactly which past terms succeeded in denoting real entities, and which did not. This conclusion seems to us to accord perfectly well with intuition about the referential success of past theoretical terms. In particular, it seems to avoid both the excessive generosity of the Kripke–Putnam account and the excessive stinginess of total-theory descriptivism.

In any case, our current concern is not to adjudicate on the extent of past referential success, but to decide whether theories are equivalent to their Ramsey-sentences. In this connection, as opposed to adjudicating past referential success, it does not matter too much exactly which assumptions in a theory go into defining its terms. For pretty much any set of assumptions in this role will have the effect of reducing a theory to its Ramsey-sentence.

To see this, return to our earlier analogy. Suppose we define 'John' as 'the (unique) tall boy'. Then the theory that 'John is tall, thin, and red-haired' will clearly imply the 'Ramsey-sentence' that 'There is a unique tall, thin, red-haired boy'.

True, there is not quite an implication the other way round. 'There is a unique tall, thin, red-haired boy' does not imply 'The (unique) tall boy is tall, thin and red-haired', for there may not be a unique tall boy, even if there is a unique tall, thin, red-haired boy.

Still, we think we can reasonably ignore this slight mismatch between a Ramsey-sentence and its theory, arising from the fact that only some of the theory may be used in defining its terms. The possibility that the Ramsey-sentence will be true, but the original theory false, will only be actualized in those cases where the Ramsey-sentence identifies a unique entity, but the descriptions used in defining the relevant term do not so ensure uniqueness. But such cases are unlikely to be actual. Scientific thinkers may not use their whole theory to define their terms. But they will normally pack enough into their descriptions to pick out a unique entity, whenever the whole theory does so. (Cf. Papineau, 1996. There may well be cases where uniqueness is not achieved. But what we doubt is that there are cases where this failure of uniqueness would be remedied by including the whole theory in place of the smaller set of defining descriptions.)

Names and descriptions

Let us now consider a different kind of reason for doubting the equivalence of theories and their Ramsey-sentences. Someone may agree that if theoretical terms were equivalent to Russellian descriptions – that is, if they were contextually defined terms to be systematically eliminated in line with Russell's theory of descriptions – then theories would indeed come to the same thing as existentially quantified Ramsey sentences. But they could object that this is not the same as viewing theoretical terms as genuine terms whose referents happen to be fixed by descriptions.

The point is that someone could accept the superiority of a descriptivist view over the Kripke–Putnam causal view, for the reasons given in the last section, and yet insist that this does not show that theoretical terms in science can be eliminated in favour of Russellian existential quantifications. For these terms may still be genuine names. The last section only shows that, if so, they are names whose reference is fixed by description, rather than determined causally *à la* Kripke and Putnam.

If this is right, the objection can then continue, scientific theories are by no means equivalent to their Ramsey-sentences, and Laudan's original objection to realism about scientific theories will stand. For a claim made using an empty genuine term is no claim at all, and a theory formulated using empty scientific terms will be similarly empty, and therefore no candidate for approximate truth.

Moreover, we have been given no reason to suppose that the terms on which Laudan focuses will succeed in referring just because they are introduced via descriptions. On the contrary, the last section advertised it as a selling point of the descriptivist view, over the Kripke–Putnam causal account, that it was not overly generous, and would deny reference to those scientific terms that intuitively seem not to have referred to anything. So it seems that, even if 'aether' has its putative reference fixed by description, it will still refer to nothing, with the result that there is no question of whether 'that entity' approximately satisfies the properties nineteenth-century electromagnetic theory 'attributed to it'.

We do not accept this objection. It seems to us to postulate too sharp a dichotomy between genuine terms and Russellian descriptions. Our view is that a scientific term introduced by a description may behave like a name in other respects, and yet the truth conditions of the sentences it enters into still conform to Russell's analysis of descriptions. Thus we accept that scientific terms may behave rigidly in modal contexts (equivalently – behave like an 'actualized' description), and that speakers may build up their understanding of sentences involving them in the same way as they build an understanding of sentences involving simple names. But we take this to be consistent with sentences involving those terms having the truth-conditional content that there exist entities with such-and-such properties, that is, being equivalent to their Ramsey-sentences (cf. Sainsbury, forthcoming). If this is right, then scientific theories will still be candidates for approximate truth, even if their terms fail to refer, since a claim that there exist entities with such-and-such properties can still be approximately true even if there are no actual entities of precisely those kinds.

There is one very strong argument in favour of this Russellian view of scientific terms – namely, the possibility of true negative existential claims. There is no phlogiston, no aether, and no natural places. Obviously true. But the view that empty scientific terms cannot be used to say anything has trouble accommodating these obvious truths. ('If there is no phlogiston, then how can you say that it doesn't exist?') Our more Russellian view has no difficulty on this point, since it equates the negative existential claims with the negation of an existential quantification. To say there is no phlogiston is simply to say that there is no entity with such-and-such properties.

Let us conclude by considering one further circumstance that might be thought to count against the descriptivist view and in favour of the non-Russellian alternative. The best examples of non-Russellian names are names whose understanding does not seem to depend on speakers' grasp of any descriptions. Ordinary proper names which work according to Kripke's causal theory of reference would be the paradigm. In such cases, there is no possibility of a quantificational Ramsey-sentence reading of claims involving the terms, for lack of any associated descriptions. And so here there is no alternative to the conclusion that claims made using empty terms fail to say anything.

Now, so far we have been taking it for granted that the people who use scientific terms will not be short of the relevant descriptions, on the grounds that these are simply the descriptions embodied in the relevant scientific theory. But on reflection this is clearly an idealization. Plenty of people are competent with scientific terms, and are held responsible for claims they make using them, even though they do not fully grasp the theory in which the terms are used. We have in mind here not only lay people, who have heard the terms in scientists' mouths and intend to use them with the same reference, but also many scientists themselves. Science is a rambling enterprise, whose practitioners include experimenters, engineers, and research students, as well as theoreticians, and by no means all will fully grasp the theories in which the terms they use are embedded.

Because of this, it could well be argued that scientific claims, as constituted in the thinking of the general run of scientists, cannot possibly be equivalent to their Ramsey-sentences, for this equivalence requires a descriptional equivalent for scientific terms, and no such equivalent is available to the general run of scientists. So, once more, it seems that scientific terms must work as simple non-Russellian names, with the result that claims made using empty terms say nothing, and so *a fortiori* cannot be approximately true, just as originally assumed by Laudan.

We accept that this may well be the right view of scientific claims as thought by the general run of scientists. Since most scientists will not have any specific descriptions in mind when they use a scientific term, their claims can only be read as about that referent, if there is one, and as saying nothing, if there is not. We cannot equate their claims with Ramsey-sentences, if they do not have the descriptions with which to formulate such Ramsey-sentences.

Still, we do not think this invalidates the position adopted in this chapter. For we do not think that scientific claims as thought by the general run of scientists are the right thing to focus on when considering how the history of science bears on scientific realism. When we look back at some past theory, with a view to considering how far it was true, it would clearly be inappropriate to conceive of it in the way it was understood by those who only had a partial grasp of the theory.

For a start, this would stop us concluding that phlogiston does not exist. (What does not exist, if we are not allowed to think of phlogiston descriptively?)

And, more fundamentally, it would be very odd to think of past theories in the way they were understood by those who did not really know them, given that our concern is precisely with how far those past theories were true. Given that we are trying to assess the approximate truth of those theories themselves, we need to focus on the full theories, as articulated by those who did understand them. And here there is no need to suppose that these people understood their scientific terms as mere names, divorced from any descriptive content. For, unlike the general run of scientists, they would have grasped the descriptions which were used to introduce the terms, and would have been able to continue understanding the terms in this descriptive way.

So when we consider past scientific theories as thought by the scientists who knew them we can still view the relevant terms as Russellian descriptions, and so view the whole theories as equivalent to their Ramsey-sentences. And if this is right, the defence of Ramsey-sentence realism offered in the earlier sections of this chapter will constitute a defence of scientific realism as normally construed.

References

Demopoulos, W. and Friedman, M. (1985), 'Critical Notice: Bertrand Russell's *The Analysis of Matter*, its Historical Context and Contemporary Interest', *Philosophy of Science*, 52, pp. 621–39.

Hardin, C. and Rosenberg, A. (1982), 'In Defense of Convergent Realism', Philosophy of Science, 49, pp. 604–15.

Kitcher, P. (1993), *The Advancement of Science*, New York, Oxford University Press.

Laudan, L. (1981), 'A Confutation of Convergent Realism', *Philosophy of Science*, 48, pp. 19–49.

Laudan, L. (1984), 'Realism without the Real', *Philosophy of Science*, 51, pp. 156–63.

Lewis, D. (1970), 'How to Define Theoretical Terms', *Journal of Philosophy*, 67, pp. 249–58.

Lewis, D. (1984), Putnam's Paradox', *Australasian Journal of Philosophy*, 62, 3, pp. 221–36.

Newton-Smith, W. (1981), *The Rationality of Science*, Boston, Routledge and Kegan Paul.

Papineau, D. (1996), 'Theory-Dependent Terms', *Philosophy of Science*, 63, pp. 1–20.

Putnam, H. (1978), *Meaning and the Moral Sciences*, London, Routledge and Kegan Paul.

Sainsbury, M. (forthcoming), 'Referring Descriptions'.

Worrall, J. (1989), 'Structural Realism: The Best of Both Worlds?', *Dialectica*, 43/1, pp. 99–125.

Chapter Ten

The Limits of Realism

Michele Marsonet

Ontology and epistemology

To what extent are we entitled to draw a borderline between ontology and epistemology? To some contemporary thinkers a positive answer to this question looks attractive, mainly because it reflects convictions deeply entrenched in our common-sense view of the world. However – they argue – anyone wishing to clarify the distinction between the ontological and the epistemological dimensions, without having recourse to unwarranted dogmas, should recognize that such an answer poses more problems than it is meant to solve. This is due to the fact that the separation between factual and conceptual is not sharp and clean, but rather fuzzy.[1] To this recognition another remark should be added. As long as humans are concerned – so the argument goes – the world is characterized by a sort of 'ontological opacity' which makes the construction of any absolute ontology very difficult. Our ontology is characterized by the fact that the things of nature are seen by us in terms of a conceptual apparatus that is inevitably influenced by mind-involving elements.[2] All this has important consequences on both the question of scientific realism and the realism–anti-realism debate.

Theoretically, we may admit that a distinction can be drawn between the natural world on the one hand, and the social-linguistic world on the other. However, it should not be difficult to understand that we began to identify ourselves and the objects that surround us only when the social-linguistic world emerged from the natural one, and this in turn means that our criteria of identification are essentially social and linguistic. Leaving aside any kind of Platonism, and recognizing – in a pragmatist vein – that the concept of 'truth' is essentially tied to human interests, we need an intersubjective criterion giving rise to the notion of a world which is both objective and mind-independent. In other words, as John Dewey stated, the distinction subject–object is not to be found in nature: it arises when men have such an intersubjective criterion, that is, within a social world which is created by men themselves.[3] But it is important to note at the onset that these remarks do not entail the total identification of the aforementioned two worlds. The conclusion is rather that, of the natural world as such, little can be said. We can admit that a borderline between ontology and epistemology really exists but, as far as we are concerned, such a distinction looks less definable today than it was usually thought to be in the past.

There are two reasons which explain why things are so. On the one hand conceptualization gives us access to the world, while, on the other, it is the most important feature of our cultural evolution (which is distinct from – although not totally alien to – biological evolution).[4] This does not mean to diminish the importance of the latter, which is specifically geared to the natural world and, after all, precedes our cultural development from the chronological viewpoint. However, it is cultural evolution that distinguishes us from all other living beings. While the idealistic thesis according to which the mind produces natural reality looks hardly tenable, it is reasonable to claim instead that we perceive this same reality by having recourse to the filter of a conceptual apparatus whose presence is, in turn, connected to the development of language and social organization.

All this prevents a clear distinction between ontology and epistemology. For example, it might be stated that ontology's task is to discover what kinds of entities make up the world ('what there is', in Quine's terms), while epistemology's job is to ascertain what are the principles by which we get to know reality. It is obvious, however, that if our conceptual apparatus is at work even when we try to pave our way towards an unconceptualized reality, our access to it inevitably entails the involvement of the mind. Resorting to a paradox, it might even be said that any unconceptualized reality turns out to be an image of the mind (even though, it is worth repeating it, this recognition does not force us to deny the mind-independent existence of unconceptualized reality).

At this point an important problem must be faced. Since the rejection of any scheme–content distinction looks hardly tenable,[5] the question arises whether it is more appropriate to speak of 'scheme' (singular) or of 'schemes' (plural). This is not a rhetorical question, as it might seem at first sight. What lies behind it is, rather, the question of ontological pluralism, which is in turn connected to the existence of possible alternative ways of conceptualizing reality.

The importance of such a question was clearly understood by William James. At the beginning of the twentieth century, in fact, he wrote:

> It is possible to imagine alternative universes to the one we know, in
> which the most various grades and types of union should be
> embodied ... we can imagine a world of things and of kinds in
> which the causal interactions with which we are so familiar should
> not exist.[6]

James went on to say:

> The 'absolutely' true, meaning what no farther experience will ever
> alter, is that ideal vanishing-point towards which we imagine that all
> our temporary truths will some day converge ... meanwhile we have
> to live to-day by what truth we can get to-day, and be ready to-
> morrow to call it falsehood.[7]

The conclusion of this line of reasoning is that the great scientific and
metaphysical theories of the past were adequate for centuries but, since
human experience has 'boiled over' those limits, we now call these theories
only relatively true. Those limits were in fact casual, and 'might have been
transcended by past theorists just as they are by present thinkers'.[8]

The American pragmatist was not the first to note that our world-view
can never be absolute, and that intelligent creatures whose experiential
modes are substantially different from our own are bound to conceptualize
reality in a rather diverse way. James, however, provided us with a clear
picture which anticipates the contemporary debate on conceptual schemes.
He claimed in this respect that:

> In practical talk, a man's common sense means his good judgement,
> his freedom from excentricity [sic] ... In philosophy it means
> something entirely different, it means his use of certain intellectual
> forms or categories of thought. Were we lobsters, or bees, it might
> be that our organization would have led to our using quite different
> modes from these of apprehending our experiences. It *might* be too
> (we can not dogmatically deny this) that such categories,
> unimaginable by us to-day, would have proved on the whole as
> serviceable for handling our experiences mentally as those which
> we actually use.[9]

Rationality and possibility

Someone might object that these are only mental experiments, whose
importance cannot be overestimated. However, mental experiments play a
key role in both philosophy and science. No doubt they are merely

hypothetical devices, but they also allow us to enter the dimension of possibility. By resorting to them, we are able to imagine how the world could have been in the past, could be today, or could turn out to be in the future. This is a specific characteristic of our relationship with the world, which is strictly connected to the cultural type of evolution mentioned above. Rationality is, thus, largely a matter of idealization. Although our natural origins and evolutionary heritage must be duly deemed important, we must also recognize that there is indeed something that makes us unique. Only human beings are able to take idealities into account and to somehow detach themselves from the actual world. Rationality may also be seen as the expression of mankind's capacity to see not only how things actually are, but also how they might have been and how they could turn out to be if we were to take some courses of action rather than others: the concept of possibility indeed plays a key role. It should eventually be noted that the dimension of possibility plays quite an important role even in the scientific domain, since scientific theories concern possible rather than actual reality. Newton's theory of universal gravitation takes into account the ideal mass in ideal space, and its status as scientific theory is established by the fact that it holds for any mass.

In short, *possibilia* are key components of our social-linguistic world, that is, of the specifically human way of dealing with reality. Possible worlds and possible individuals are actual or potential products of our conceptual apparatus, and any strategy meant at eliminating them appears doomed to failure. The dimension of possibility, besides being strictly tied to hypothetical reasoning, plays a fundamental role in our comprehension of both the natural and social-linguistic worlds. But it should also be clear that such a dimension must anyhow make reference to some kind of agent, and the agent itself is thus an inevitable point of departure. We are compelled to adopt such a stance, because this is the only way open to us for gaining access to the world. No one denies that it would be good to transcend our conceptual machinery in order to glimpse how the world really is, independently of any view we can hold about it. This, however, cannot be done because of the very way we are made. Unlike some forms of classical idealism, we can recognize the presence of things that are 'real' in the sense of being mind-independent but, on the other hand, a qualification is needed to the effect that human beings have access to those things only via their conceptual apparatus.

Starting from such premises, it is reasonable to claim that (1) analytic and synthetic cannot be clearly separated, and (2) no neatly determinable distinction can be drawn between science and metaphysics. As Quine stated in the 1950s:

The totality of our so-called knowledge or beliefs, from the most casual matters of geography and history to the profoundest laws of atomic physics or even of pure mathematics and logic, is a man-made fabric which impinges on experience only along the edges. Or, to change the figure, total science is like a field of force whose boundary conditions are experience ... Revision even of the logical law of the excluded middle has been proposed as a means of simplifying quantum mechanics; and what difference is there in principle between such a shift and the shift whereby Kepler superseded Ptolemy, or Einstein Newton, or Darwin Aristotle?[10]

Science and common sense

A follower of scientism might at this point be tempted to affirm the unconditioned superiority of the scientific world-view over the image of the world that Wilfrid Sellars used to call the 'manifest image', that is the common-sense image which is shared – in its broad features – by all men *qua* men.[11] But is it really plausible to claim that science deserves the primary role in assessing any kind of conceptual scheme? What guarantees can science provide in this regard? And, above all, which science are we talking about in this context? No doubt the real world contains those entities which would be posited by an 'ideally complete' science such as the one envisioned by Charles S. Peirce. But this ideal completeness is not available, and we are therefore compelled to work with what we have at our disposal now. This takes us back to the current scientific world-view, that is to say, the one provided by today's science. We must face, in sum, a notion of truth which is essentially 'relative' and bound to evolve with the passing of time.

In other words, the presence of a sort of Peircean ideal community of scientific researchers who are supposedly able to attain the 'real truth' about the world is not an option, but an indispensable condition for the truthfulness of our generalizations about reality. Peirce, in fact, made clear that the key characteristic of truth is stability, and that a true belief must at least be fated to be underwritten by the operation of scientific method.[12] Of course we cannot rule out the possibility that such an ideal community will exist in the future, but the history of science should at least prompt us to be pessimistic in this regard. Ideal science, even when its realization is referred to the future, looks more a philosophical utopia than a feasible accomplishment (even though, as is well known, utopias are indeed useful when they are viewed as essentially 'regulative' ideas). The strong realistic thesis that science faithfully describes the real world turns out to be, thus, just a matter of intent.

The fact is that scientific world-views continuously evolve, which means that the scientific enterprise has an essentially historical character. As Werner Heisenberg pointed out, science is always the result of the encounter between the natural world on the one side, and human conceptions, practical interests, and needs on the other.[13] The appeal to mental experiments is useful not only in daily life, but in the scientific domain too, because in this case science itself makes clear that it permits us to know the world from a particular perspective, that is in turn geared to the specific relationships we entertain with the environment which surrounds us. John Dewey used the term 'transaction' to denote this encounter, where the respective contributions of the observer and of the observed reality cannot be rigidly distinguished.[14]

This means that our science is essentially relational, and not absolute. The information with which it provides us is appropriate, but from our point of view. The Jamesian point that it is possible to imagine alternative universes to the one we know, and that intelligent creatures whose experiential modes are substantially different from our own are bound to interpret reality in a different way, must be taken seriously. In other words, we should recognize that the natural environment in which we live (and of which we are a substantial part) has an essential bearing on conceptualization, including the scientific one. Science provides reliable information about the world, but this information is always relative to a particular conceptual framework, and it is a mistake to think that the limits of our cognitive capacities have only an aprioristic character. We are mainly bound by empirical limits, due to the fact that human beings inquire into nature by means of an apparatus which answers to certain stimuli, but not others. However, nothing in our actual science leads us to rule out the hypothesis that, in other natural environments, the development of science might have taken quite a different course.[15] In order to give plausibility to this hypothesis, we must only admit the existence of worlds whose natural environment is substantially different from our own, and certainly this is not mere science fiction.

By saying this, we leave the domain of mental experiments to enter that of hypotheses which are – at least in principle – empirically verifiable. No doubt our science today is the only science we know, but this should not lead us to exclude the possibility that there are other ways of investigating nature. After all, science tells us that there are many aspects of reality that we cannot get in touch with by means of our sensory apparatus (which is the product of a process of evolution which took place in particular environmental conditions). Therefore we should not uncritically accept Davidson's statement that 'since there is at most one world, these pluralities are metaphorical or merely imagined'.[16]

Which ontology?

The question now is the following: are we authorized to claim than any absolutely objective ontology should be left in the background, because little can be known about it? It should be noted that not only philosophers, but even many scientists, have often denied the validity of the general picture of the world that the man in the street takes more or less for granted. In our century uncertainty about the content of our theories has grown fast, together with the feeling that there are alternative theories that can account equally well for all possible observations. Clearly the threat of relativism arises at this point, even though many authors nowadays no longer take relativism to be a threat, but just a matter of fact.

All this explains why the issue of conceptual schemes is important for both philosophers and scientists. For example, according to Niels Bohr's principle of complementarity we have, on the one side, a sort of Kantian world-in-itself which is both unknowable and undescribable, and on the other an 'us' which, unlike in Kant's picture, is not stable and determined. This means that, in our inquiries about the world, different questions can all receive coherent answers, with the disquieting effect that a comprehensive and coherent image of reality cannot be achieved. It is as if, conducting different experiments, we were to change conceptual scheme: the world experienced will in any case be diverse, and there is no way to combine the world of our experience with the various, differing conceptual schemes. The peculiar form of quantum effects entails that ordinary classical ideas about the nature of the physical world are profoundly incorrect, and some contemporary physicists endorse in this respect views which recall William James's characterization of consciousness as a 'selecting agency'.[17]

Obviously things were different when logical positivism was still the dominant doctrine in the philosophy of science. In that case the main purpose was to individuate the immutable models that lie beyond concrete scientific practice, because it was commonly held that science is objective and progressive in the cumulative sense of the term. It must be stressed, however, that distance from the neo-positivist model does not lead one automatically closer to some kind of methodological anarchism or postmodernism (in Rorty's sense of the term). Some authors, in fact, claim that science can effectively validate a plausible commitment to the actual existence of its theoretical entities. But scientific conceptions can get, at most, a rough consonance between our scientific ideas and reality.[18] And this statement should not sound surprising, if only one recalls what we said before about the difficulty of tracing a precise borderline between ontology and epistemology.

The general picture that emerges from the preceding remarks is the following. It is certainly correct to state that science aims to offer correct information about the world, but the extent to which it succeeds in accomplishing this task is always questionable. We cannot claim that the picture provided by today's science – our current scientific image of the world – is absolutely correct, because the history of science shows that any such statement is likely to be rejected by future generations. While it may be recognized that science purports to offer a correct description of the real world, past experience should also prompt us to accept its claims *sub condicione*, and to view them as merely provisional.

It has often been said in this regard that the actual unobservability of scientific entities rests on contingent facts, which depend on both the nature of the unobserved thing and features of our perceptual mechanisms. This means that things which were in the past unobservable became observable later on, because we were able to artificially extend our perceptual capacities by means of technologically advanced scientific instruments. By accepting these premisses, any neat demarcation between observable and unobservable entities is not significant from an ontological point of view. It should be noted, however, that if we reject the realist perspective as far as scientific unobservable entities are concerned, even realism in general must be abandoned. Following this line of thought unobservable scientific entities are just contingently unobservable, so that their unobservability (due, for instance, to smallness of size) presents the same, resolvable difficulties that one has to deal with when far distant celestial bodies are taken into account (in the latter case, spatial location is the problem at issue).

The preceding arguments may be accepted with some reservations. Clearly, we must be ready to admit the reality of the so-called theoretical entities if we want to avoid an instrumentalistic conception of scientific knowledge. After all, it is easy to verify that scientists deem prediction and control important just because they are supposed to monitor the adequacy of our scientific theories about reality. This is reason why anti-realism has never been popular among scientists. As the physicist Steven Weinberg has it:

> The insights of philosophers have occasionally benefited physicists, but generally in a negative fashion – by protecting them from the preoccupations of other philosophers ... Physicists do of course carry around with them a working philosophy. For most of us, it is a rough-and-ready realism, a belief in the objective reality of the ingredients of our scientific theories.[19]

In other words, while it is correct to state the fallibility and continuous corrigibility of science, starting from these premises we are not allowed to draw the conclusion that no existential and descriptive claims about the real world should be made in scientific theorizing.

Science constantly attempts to provide answers to our questions about how things stand in the world, and thus purports to offer reliable information about it. But it should also be recognized that the extent to which science succeeds in accomplishing this task is disputable. No doubt relativity theory and quantum mechanics are the best scientific theories we have at our disposal now, but to assume that they will still be deemed adequate in the future is rather dangerous. Even theoretical entities of science are introduced for a utilitarian purpose, that is, to provide the materials of causal explanation for the behaviour of real things. This means that science is not a merely practical instrument for prediction and control that has no bearing on describing the nature of the world. Our science's claims regard the real world, but they are always tentative.

All we are entitled to say is that if our present science is correct, then the so-called theoretical entities exist and possess the characteristic features that it envisions. No science would indeed be possible without this basic realistic attitude, because its very aim is to provide an ontologically founded picture of reality. In understanding this fact, a philosopher of science has to recognize, on the one side, the descriptive and explanatory role that science purports to play, while, on the other, he must also stress that science is bound to be imperfect and fallible in its execution of such a role.

Strong and weak realism

At this point we are confronted by a crucial point: what kind of realism – if any – can we actually endorse? The question becomes even more important if we recall that, for the reasons stated above, many authors claim that no border between ontology and epistemology can be outlined. It is often stated that, in order to provide realism with a solid foundation, we need recourse to a reality that is totally independent of thought (and let alone of language).[20] We should, therefore, ask ourselves: What can we possibly think about this reality, and how can we say what it is like? Even when we imagine a world totally devoid of human presence, we must use human concepts. As we noted previously, conceptualization is not an option we can get rid of, but a built-in component of the nature of human beings.

Is it true, however, that the aforementioned claim heavily relies on the presence of an alleged capacity to get a view of the world which is totally independent from the experiencing subject? As is well known, such a thesis has constantly been rejected by the pragmatist tradition. Writing about Russell's and Dewey's divergent opinions about logic, Tom Burke has made the following remarks:

> For both Dewey and Russell, a certain amount of conceptual stage setting has to go on prior to presenting a semantic theory ... For Russell, we have to be able to make certain assumptions about the world independent of our experience of it. The world is in this view carved up into objects having properties and standing in relations, and we have only to open our eyes to note such facts ... For Dewey, one jumps the gun by a long shot by making certain independent and sweeping claims about the world in this manner. In taking this stance, one is less than a step away from embracing a view referred to by Dewey as the 'spectator theory of knowledge', namely, the view that we can say something about the world (as it 'really' is) independent of our participation in it. Russell commits himself, qua logician, to such a view to the extent that he assumes a world full of facts without questioning how we come to grips with such facts in our experience, opting to focus solely on the abstract study of propositional or linguistic systems.[21]

In the philosophy of science, this means that we can never assume that a particular scientific theory gives us the true picture of reality. The current state of scientific knowledge is one among other cognitive states that share the same imperfection. There is indeed a strong prospect that many or most of our current scientific theories will be recognized to be inadequate: our current scientific knowledge is a set of hypotheses, many or most of which are likely to be regarded as untenable in the future. Not only are we not in a position to claim that our knowledge of reality is complete; we are not even in a position to claim that our knowledge of reality is correct. We need a more modest view in this case. Just as we think now that our predecessors held an inadequate vision of the world, so it is reasonable to assume that our successors will hold the same opinion about our vision of the world.

Science, in sum, is not a stable system, but a dynamic process, and this fact leads to view as problematic all those conceptions that place on the shoulders of future science the burden of perfection. Science is not rational because it has a solid foundation, but because it is a continuously evolving self-correcting enterprise, whose claims are always open to the possibility of revision. For this reason it is better to endorse a modest realism. This is the reason why the history of science plays a key role, and in this regard we think that something more can be said. Even the history of the philosophy of science is important, since it makes us understand how the models by which philosophers interpret science (and reflect on it) change.[22] We should be sceptical about any proposal which aims at distinguishing in a rigid manner science from metaphysics. There is no atemporal 'scientific image of the world', but many images located in the flux of time. The very image of common sense which – apparently at least – is quite stable, continuously evolves and incorporates elements coming from the various scientific images.

If we claim that the science of our day provides the true picture of how the world really is, we seriously risk (given the concrete situation in which we happen to live) hypostatizing something that is simply a contingent and historically determined product of our action. This product is valid in a particular period of our cultural evolution, and an approach such as the one previously envisioned should prevent us from claiming that the ontology of contemporary science is the absolute ontology that so many metaphysicians were looking for in the past. The preceding remarks prompt us to conclude that relativism and fallibilism are not ghosts to be afraid of, but just inevitable factors of our relationship with the surrounding environment. Richard Rorty is right when he notes that natural science is not a natural kind,[23] since it is essentially geared to historical and cognitive values.

Minimal realism

However, it is essential to note that the aforementioned remarks do not necessarily lead towards some form of anti-realism. It is correct to state that, due to our cognitive position in the world and its limitations, the perspective provided by the conceptual framework we employ cannot be transcended. This amounts to saying that, although the world does not need our participation in order to be, our epistemic access to the world is given by such participation. Any description, thus, is bound to be determined by our operational perspectives. We do better not to say much about an absolute reality, even though we may push our imagination so far as to imagine how it could be. Our ontology is always bound to have epistemological commitments or, to put it in different terms, ontological commitments cannot be denied an epistemological aspect.

In order to obtain final and metaphysically powerful answers we would have to detach ourselves from contingency, but we will never be able to do this. We get to know the natural environment by using scientific instruments and formulating scientific theories, but the history of the natural world with which we are acquainted is always a history that refers to human beings, because we develop it by having recourse to our conceptual apparatus. We can imagine a world in which no conceptual scheme is at work thanks to our capacity to detach ourselves from our actual situation or world, and envisage things being ways they could not actually be.

On the other hand, however, nothing prevents us from claiming that absolute reality – that is, a reality which does not depend on our cognitive capabilities – is there. After all science is important not because it provides us with the correct (that is, unique) paradigm of knowledge. It is important, rather, because it makes us understand that the world is – or might be – different from how we see it. And this, once again, casts doubts on a strange feature of Davidson's philosophy: he does not take into account the possibility that reality might be different from what we take it to be. We can admit that common sense is a sort of background comprehensive theory which grounds everything else including science: it is just our standard way of viewing the world and of dealing with it. However, we should also recognize that it possesses a practical primacy, and not an ontological one.

Even admitting that our ontology depends on communication, and that only communication allows us to hold the concept of objective truth,[24] we are not entitled to claim – as Davidson does – that our view of the world is, in its plainest features, largely correct. Starting from these premisses, we should instead say that the knowledge of our world (the common-sense or manifest image, to use once again Sellars's terms), and not the knowledge of the world as such, is largely correct. Sellars pointed out, in fact, that

> since this image [the manifest or common sense image of the world] has a being which transcends the individual thinker, there is truth and error with respect to it, even though the image itself might have to be rejected, in the last analysis, as false.[25]

As a matter of fact science introduces us to whole dimensions of reality which were previously unknown to human beings. This shows that (1) reality-as-such and reality-as-known-by-us do not always coincide, and that (2) a distinction between ontology and epistemology can – and indeed must – be made. Given our cognitive limits such a distinction is certainly hard to draw, but still, its philosophical importance is so evident that any attempt to overcome it leads to the unjustified thesis that no reality lies beyond our cognitive capabilities.

Quine seems – sometimes, at least – aware of this fact. In an essay entitled 'Existence' he makes the following claims:

Which ontology to ascribe to a man depends on what he does or intends with his variables and quantifiers. This second appeal to language is no more to be wondered at than the first; for what is in question in both cases is not just what there really is, but what someone says or implies that there is. Nowhere in all this should there be any suggestion that what there is depends on language ... It may in this sense be said that ontological questions are parochial to our culture. This is not to say that a thing may exist for one culture and be non-existent for another. Existence is absolute, and those who talk of existence can say so. What is parochial is the talking of it.[26]

What prevents him from fully drawing the proper consequences of these remarks is his thesis that ontology is totally relative to language. Quine is often led to blur any distinction between ontology and epistemology because he tends to use interchangeably two different meanings of the term 'ontology': (1) what there is and (2) what we 'believe' there is.

In the last analysis we would like to stress that, despite what many relativists claim, realism in its minimal version endorsed previously still is an arguable position. Realism is certainly an unpopular stance today but, for the reasons stated above, the standard arguments against it are by no means conclusive. And, if one asks what difference is made to our knowledge claims if we accept the existence of an extra-conceptual world, the answer is the following: such recognition is likely to undermine the widespread anthropocentric stance which identifies reality with our (limited) knowledge of it.

Notes

1   It goes without saying that the seminal work here is still Quine (1980). For a more recent perspective see McDowell (1994).
2   A view of this kind is endorsed in Rescher (1992).
3   See especially Dewey (1994). Davidson exploits Dewey's insight in Davidson (1990).
4   The distinction biological—cultural evolution is constantly present in pragmatist authors like James, Peirce, and Dewey. For a contemporary assessment see Rescher (1990).
5   See especially Davidson (1985), and Rorty (1982). We cannot take this problem into account here for reasons of space. For a recent criticism of Davidson's and Rorty's positions see Haack (1993).
6   James (1907, pp. 156–7).
7   Ibid., pp. 222–3.
8   Ibid.
9   Ibid., p. 171.
10 Quine (1980, pp. 42–3).

11 See Sellars (1963).
12 For a good analysis of this point see Rescher (1978).
13 Heisenberg (2000).
14 Dewey (1994).
15 Interesting remarks on this topic can be found in Rescher (1984).
16 Davidson (1985, p. 187).
17 See, for example, Stapp (1993).
18 Such a stance is defended in Rescher (1987).
19 Weinberg (1992, pp. 166–7).
20 In other words, we should adopt what Putnam calls the 'God's Eye point of view'; see Putnam (1981).
21 Burke (1994, pp. 56–7).
22 See, for instance, Oldroyd (1986).
23 Rorty (1991).
24 See Davidson (1996).
25 Sellars (1963, p. 14).
26 Quine (1970, pp. 93–4).

References

Burke, T. (1994), *Dewey's New Logic: A Reply to Russell*, Chicago, University of Chicago Press.
Davidson, D. (1985), 'On the Very Idea of a Conceptual Scheme', in *Inquiries into Truth and Interpretation*, Oxford, Clarendon Press, 1985, pp. 183–98.
Davidson, D. (1990), 'The Structure and Content of Truth', *Journal of Philosophy*, 57, 1990, pp. 279–328.
Davidson, D. (1996), 'Subjective, Intersubjective, Objective', in P. Coates and D. D. Hutto (eds), *Current Issues in Idealism*, Bristol, Thoemmes Press, 1996, pp. 155–77.
Devitt, M. (1991), *Realism and Truth*, Oxford, Blackwell, 2nd ed.
Dewey, J. (1994), *Experience and Nature*, Chicago-La Salle, Illinois. Open Court, 9th pr.
Haack, S. (1993), *Evidence and Inquiry. Towards Reconstruction in Epistemology*, Oxford, Blackwell.
Heisenberg, W. (2000), *Physics and Philosophy*, London, Penguin.
James, W. (1907), *Pragmatism*, London/New York, Longmans, Green and Co.
McDowell, J. (1994), *Mind and World*, Cambridge, MA, Harvard University Press.
Oldroyd, D. (1986), *The Arch of Knowledge. An Introductory Study of the History, Philosophy, and Methodology of Science*, New York, Methuen.
Putnam, H. (1981), *Reason, Truth and History*, Cambridge, Cambridge University Press.
Quine, W. V. O. (1970), 'Existence', in W. Y. Yourgrau, A. D. Breck (eds), *Physics, Logic and History*, New York, Plenum Press, 1970, pp. 89–103.
Quine, W. V. O. (1980), 'Two Dogmas of Empiricism', in W. V. O. Quine, *From a Logical Point of View*, Cambridge, MA, Harvard University Press, 1980, 4th pr., pp. 20–46.
Rescher, N. (1978), *Peirce's Philosophy of Science. Critical Studies in His Theory of Induction and Scientific Method*, Notre Dame/London, University of Notre Dame Press.

Michele Marsonet

Rescher, N. (1984), *The Limits of Science*, Berkeley/Los Angeles/London, University of California Press.

Rescher, N. (1987), *Scientific Realism. A Critical Reappraisal*, Dordrecht/Boston, Reidel.

Rescher, N. (1990), *A Useful Inheritance: Evolutionary Aspects of the Theory of Knowledge*, Rowman and Littlefield, Savage.

Rescher, N. (1992), *A System of Pragmatic Idealism (Vol. 1: Human Knowledge in Idealistic Perspective)*, Princeton, Princeton University Press.

Rorty, R. (1982), 'The World Well Lost', in R. Rorty, *Consequences of Pragmatism*, University of Minnesota Press, Minneapolis, 1982, pp. 3–18.

Rorty, R. (1991), 'Is Natural Science a Natural Kind?', in R. Rorty, *Objectivity, Relativism, and Truth, Philosophical Papers* (vol. 1), Cambridge, Cambridge University Press, 1991, pp. 46–62.

Sellars, W. (1963), 'Philosophy and the Scientific Image of Man', in *Science, Perception and Reality*, London, Routledge and Kegan Paul, 1963, pp. 1–40.

Stapp, H. (1993), *Mind, Matter, and Quantum Mechanics*, Berlin/Heidelberg/New York, Springer-Verlag.

Weinberg, S. (1992), *Dreams of a Final Theory*, New York, Pantheon Books.

Bibliography

Agazzi, E. (1975), 'De la théorie électroélastique à la théorie électromagnétique du champ', *Dialectica*, 29/2–3, pp. 105–26.

Agazzi, E. (1994), 'Was Galileo a Realist?', *Physis*, 31/1, pp. 273–96.

Agazzi, E. (1997), 'On the Criteria for Establishing the Ontological Status of Different Entities', in *Realism and Quantum Physics*, E. Agazzi (ed.), Amsterdam/Atlanta, Rodopi, pp. 40–73.

Berkeley, George (1710), *Principles of Human Knowledge*.

Blackburn, Simon (1990), 'Filling in Space'. Reprinted from *Analysis* in S. Blackburn, *Essays in Quasi-Realism*, Oxford, Oxford University Press, 1993, pp. 255–8.

Boyd, Richard N. (1984), 'The Current Status of Scientific Realism', in Leplin 1984a, pp. 41–82.

Broad, Charlie D. (1978), *Kant: An Introduction*, Cambridge, Cambridge University Press.

Brown, James Robert (1994), *Smoke and Mirrors: How Science Reflects Reality*, New York, Routledge and Kegan Paul.

Burke, T. (1994), *Dewey's New Logic: A Reply to Russell*, Chicago, University of Chicago Press.

Carnap, Rudolf (1950), *Logical Foundations of Probability*, Chicago, University of Chicago Press.

Cassirer, Ernst (1922) [1907, 1911], *Das Erkenntnisproblem in der Philosophie und Wissenschaft der neueren Zeit*, vol. 2, 3rd ed., Berlin, Cassirer.

Cassirer, Ernst (1953) [1910], *Substance and Function and Einstein's Theory of Relativity* (1923) [*Substanzbegriff und Funktionsbegriff* (1910) and *Zur Einstein'schen Relativitätstheorie* (1921)], authorized English translation by W. Curtis Swabey and M. Collins Swabey, New York, Dover.

Cassirer, Ernst (1981) [1918], *Kant's Life and Thought* [*Kant's Leben und Lehre*], English translation by J. Haden, Introduction by S. Körner, New Haven, Yale University Press.

Churchland, P. M. (1979), *Scientific Realism and the Plasticity of Mind*, Cambridge, Cambridge University Press.

Churchland, P. M. (1988), *Matter and Consciousness. A Contemporary Introduction to the Philosophy of Mind*, Cambridge, MA/ London, MIT Press, revised ed.

Davidson, D. (1985), 'On the Very Idea of a Conceptual Scheme', in D. Davidson, *Inquiries into Truth and Interpretation*, Oxford, Clarendon Press, 1985, pp. 183–98.

Davidson, D. (1990), 'The Structure and Content of Truth', *Journal of Philosophy*, 57, 1990, pp. 279–328.

Davidson, D. (1991), 'Three Varieties of Knowledge', in A. Phillips Griffiths (ed.), *A. J. Ayer: Memorial Essays*, Cambridge, Cambridge University Press, 1991, pp. 153–66.

Davidson, D. (1996), 'Subjective, Intersubjective, Objective', in P. Coates and D. D. Hutto (eds), *Current Issues in Idealism*, Bristol, Thoemmes Press, 1996, pp. 155–77.

Day, Timothy and Kincaid, Harold (1994), 'Putting Inference to the Best Explanation in its Place', *Synthese*, 98, pp. 271–95.

Demopoulos, W. and Friedman, M. (1985), 'Critical Notice: Bertrand Russell's *The Analysis of Matter*, its Historical Context and Contemporary Interest', *Philosophy of Science*, 52, pp. 621–39.

Descartes, René (1641), *Meditations on First Philosophy*.

Devitt, M. (1984), *Realism and Truth*, Oxford, Blackwell.

Devitt, M. (1991), 'Aberrations of the Realism Debate', *Philosophical Studies*, 61, 1991, pp. 43–63.

Devitt, M. (1996), *Coming to Our Senses: A Naturalistic Defense of Semantic Localism*, New York, Cambridge University Press.

Devitt, M. (1997), *Realism and Truth*, Princeton, Princeton University Press, 2nd ed. with a new Afterword (1st ed. 1984, 2nd ed. 1991).

Devitt, M. (1998), 'Naturalism and the A Priori', *Philosophical Studies*, 92, pp. 45–65.

Dewey, J. (1994), *Experience and Nature*, Chicago/La Salle, Illinois. Open Court, 9th pr.

Doppelt, Gerald (1990), 'The Naturalist Conception of Methodological Standards in Science: A Critique', *Philosophy of Science*, 57, pp. 1–19.

Dowe, Phil (1992), 'Wesley Salmon's Process Theory of Causality and the Conserved Quantity Theory', *Philosophy of Science*, 59, pp. 195–216.

Dowe, Phil (2000), *Physical Causation*, Cambridge, Cambridge University Press.

Dretske, Fred I. (1981), *Knowledge and the Flow of Information*, Cambridge, MA, MIT Press.

Duhem, Pierre (1962), *The Aim and Structure of Physical Theory* [*La Théorie Physique: Son Objet, Sa Structure* (1904–6, 1914)], English translation by P. P. Wiener, from the 2nd ed. (1914), New York, Atheneum.

Dummett, M. (1978), *Truth and Other Enigmas*, Cambridge, MA, Harvard University Press.

Dummett, M. (1991), *The Logical Basis of Metaphysics*, Cambridge, MA, Harvard University Press.

Ellis, Brian (1979), *Rational Belief Systems*, Oxford, Blackwell.

Ellis, Brian (1990), *Truth and Objectivity*, Oxford, Blackwell.

Fales, Evan (1988), 'How to be a Metaphysical Realist', in *Midwest Studies in Philosophy, Volume XII: Realism and Antirealism*, Peter A. French, Theordore E. Uehling, Jr., and Howard K. Wettstein (eds), Minneapolis, University of Minnesota Press, pp. 253–74.

Fales, Evan (1990), *Causation and Universals*, New York/London, Routledge and Kegan Paul.

Feldman, F. (1974), 'Kripke on the Identity Theory', *Journal of Philosophy*, 71, p. 665–76.

Feldman, F. (1992), *Confrontations with the Reaper*, New York, Oxford University Press.

Feyerabend, Paul (1975), *Against Method*, London, New Left Books.

Feyerabend, Paul K. (1987), 'Consolations for the Specialist' (1970), in Lakatos and Musgrave (eds), pp. 197–230, 1987.

Feyerabend, P. K. (1995), 'Reply to Criticism: Comments on Smart, Sellars and Putnam', in P. K. Feyerabend, *Realism, Rationalism and Scientific Method, Philosophical Papers*, vol. 1, Cambridge, Cambridge University Press, 1995, pp. 104–31.

Field, Hartry (1978), 'Mental Representation', *Erkenntnis*, 13, pp. 9–61.

Field, Hartry (1998), 'Epistemological Nonfactualism and the A Prioricity of Logic', *Philosophical Studies*, 92, pp. 1–24.

Fine, Arthur (1984), 'The Natural Ontological Attitude', in *Scientific Realism*, Jarrett Leplin (ed.), pp. 83–107, 1984.

Fine, Arthur (1986a), *The Shaky Game: Einstein, Realism, and the Quantum Theory*, Chicago, University of Chicago Press.

Fine, Arthur (1986b), 'Unnatural Attitudes: Realist and Instrumentalist Attachments to Science', *Mind*, 95, pp. 149–77.

Fleck, Ludwik (1979) [1935], *Genesis and Development of a Scientific Fact* [*Entstehung und Entwicklung einer wissenschaftlichen Tatsache. Einführung in die Lehre vom Denkstil und Denkkollektiv*], T. J. Trenn and R. K. Merton (eds), English translation by F. Bradley and T. J. Trenn, Chicago/London, University of Chicago Press.

Galileo, G. (1929–39), *Opere*, Edizione Nazionale, Firenze, Barbera, 20 vols.

Gasking, Douglas (1955), 'Causation and Recipes', *Mind*, 64, pp. 479–87.

Goodman, Nelson (1978), *Ways of Worldmaking*, Indianapolis, Hackett Publishing Company.

Grayling, A. C. (1997), *An Introduction to Philosophical Logic*, Blackwell, Oxford, 3rd ed.

Gupta, Anil (1980), *The Logic of Common Nouns*, New Haven, Yale University Press.

Haack, Susan (1987), 'Realism', *Synthese*, 73, pp. 275–99.

Haack, Susan (1993), *Evidence and Inquiry: Towards Reconstruction in Epistemology*, Oxford, Blackwell.

Hacking, Ian (1983), *Representing and Intervening: Introductory Topics in the Philosophy of Natural Science*, Cambridge, Cambridge University Press.

Hardin, C. and Rosenberg, A. (1982), 'In Defense of Convergent Realism', *Philosophy of Science*, 49, pp. 604–15.

Heisenberg, W. (2000), *Physics and Philosophy*, London, Penguin.

Hempel, Carl Gustav (1988), 'Limits of a Deductive Construal of the Function of Scientific Theories', in *Science in Reflection*, 'The Israel Colloquium: Studies in History, Philosophy, and Sociology of Science', Ullmann-Margalit (ed.), vol. 3, Dordrecht, Kluwer, pp. 1–15.

Hempel, Carl Gustav (1992), 'Eino Kaila and Logical Empiricism', in I. Niiniluoto, M. Sintonen, G. H. Von Wright (eds), *Eino Kaila and Logical Empiricism*, 'Acta Philosophica Fennica', 52, pp. 43–51.

Hesse, Mary (1967), 'Laws and Theories', in *The Encylopedia of Philosophy*, Paul Edwards (ed.), New York, Macmillan, vol. 4, pp. 404–10.

Hick, J. (1990), *Philosophy of Religion*, Englewood Cliffs, NJ, Prentice-Hall, 4th ed.

Hitchcock, Christopher Read (1993), 'A Generalized Probabilistic Theory of Causal Relevance', *Synthese*, 97, pp. 335–64.

Hooker, Clifford A. (1974), 'Systematic Realism', *Synthese*, 51, pp. 409–97.

Horwich, Paul (ed.) (1994), *Theories of Truth*, Aldershot, Dartmouth.

Hughes, C. (2001), 'Identità personale ed entità personale' in A. Bottani and N. Vassallo (eds), *Identità personale. Un dibattito aperto*, Naples, Loffredo, 2001, pp. 157–97.

Hume, David (1888) [1739–40], *A Treatise of Human Nature*, L. A. Selby-Bigge (ed.), Oxford, Clarendon Press.

Hume, David (1955) [1740], 'Abstract of *A Treatise of Human Nature*', in C. W. Hendel (ed.).

Hume, David (1955) [1748], *An Inquiry Concerning Human Understanding*, C. W. Hendel (ed.), Indianapolis, Bobbs-Merrill. [Modern spelling of 'Enquiry' introduced by the editor.]

Hume, David (1955) [1776], 'My Own Life', C. W. Hendel (ed.), Indianapolis, Bobbs-Merrill.

Hutto, D. D. (1999), *The Presence of Mind*, Amsterdam/Philadelphia, John Benjamins.

James, W. (1907), *Pragmatism*, London/New York, Longmans, Green and Co.

Jennings, Richard (1989), 'Scientific Quasi-Realism', *Mind*, 98, pp. 223–45.

Kant, Immanuel (1985) [1781, 1787] [1929, 1933] *Critique of Pure Reason*, [*Kritik der Reinen Vernunft*], English translation by N. Kemp Smith, London, Macmillan.

Kant, Immanuel (1783), *Prolegomena to Any Future Metaphysics*.

Keynes, J. M. (1952), *A Treatise on Probability*, London, Macmillan and Co.

Kitcher, Philip (1993), *The Advancement of Science: Science Without Legend, Objectivity Without Illusions*, New York, Oxford University Press.

Kornblith, Hilary (1993), *Inductive Inference and its Natural Ground*, Cambridge, MA, MIT Press.

Kripke, S. (1971), 'Identity and Necessity', in *Identity and Individuation*, M. Munitz (ed.), New York, New York University Press.

Kripke, S. (1980), *Naming and Necessity*, Cambridge, MA, Harvard University Press.

Kuhn, Thomas S. (1962), *The Structure of Scientific Revolutions*, Chicago, Chicago University Press, 2nd ed. 1970.

Kuhn, Thomas S. (1987), 'Reflections on My Critics' (1970), in Lakatos and Musgrave (eds), pp. 231–78, 1987.

Lakatos, Imre (1987), 'Falsification and the Methodology of Scientific Research Programmes' (improved version of a paper published in 1968), in Lakatos and Musgrave (eds), pp. 91–196, 1987.

Lakatos, Imre and Musgrave, Alan (eds), (1987) [1970], *Criticism and the Growth of Knowledge. Proceedings of the International Colloquium in the Philosophy of Science*, London 1965, vol. 4, Cambridge, Cambridge University Press.

Laudan, Larry (1981), 'A Confutation of Convergent Realism', *Philosophy of Science*, 48, pp. 19–49. Reprinted in Leplin 1984a.

Laudan, Larry (1984a), 'Realism without the Real', *Philosophy of Science*, 51, pp. 156–63.

Laudan, Larry (1984b), *Science and Values*, Berkeley, University of California Press.

Laudan, Larry (1996), *Beyond Positivism and Relativism*, Boulder, Westview Press.

Leeds, Stephen (1978), 'Theories of Reference and Truth', *Erkenntnis*, 13, pp. 111–29.

Leplin, Jarrett (ed.) (1984a), *Scientific Realism*, Berkeley/Los Angeles, University of California Press.

Leplin, Jarrett (1984b), 'Introduction', in Leplin 1984a, pp. 1–7.

Lewis, D. (1970), 'How to Define Theoretical Terms', *Journal of Philosophy*, 67, pp. 249–58.

Lewis, D. (1983), 'Counterparts of Persons and Their Bodies', *Philosophical Papers*, vol. 1.

Lewis, D. (1984), Putnam's Paradox', *Australasian Journal of Philosophy*, 62, 3, pp. 221–36.

Locke, John (1690), *An Essay Concerning Human Understanding*.

Lowe, E. J. (1991), 'Persons as a Substantial Kind' in *Human Beings*, D. Cockburn (ed.), Cambridge, Cambridge University Press.

Luntley, Michael (1988), *Language, Logic and Experience: The Case for Anti-Realism*, La Salle, Open Court.

Lyotard, Jean-François (1992) [1988], *Peregrinazioni. Legge, forma, evento* [*Peregrinations. Law, Form, Event*], Bologna, Il Mulino.

Mach, E. (1883), *Die Mechanik in ihrer Entwicklung historisch-kritisch dargestellt*, Wien.

Mackie, J. L. (1973), *Truth, Probability and Paradox*, Oxford, Clarendon Press.

Mackie, J. L. (1974), *The Cement of the Universe*, Oxford, Clarendon Press.

Matheson, Carl (1989), 'Is the Naturalist Really Naturally a Realist?', *Mind*, 98, pp. 247–58.

Maxwell, J. C. (1972), *Trattato di elettricità e magnetismo*, Introduction, notes and translation by E. Agazzi, Torino, UTET, 2 vols.

McDowell, J. (1994), *Mind and World*, Cambridge, MA, Harvard University Press.

McLaren, A. (1986), 'Prelude to Embryogenesis', in *Human Embryo Research: Yes or No?*, London, Tavistock.

McMullin, Ernan (1984), 'A Case for Scientific Realism', in Leplin (1984), pp. 8–40.

McMullin, Ernan (1987), 'Explanatory Success and the Truth of Theory', *Scientific Inquiry in Philosophical Perspective*, Nicholas Rescher (ed.), University Press of America, pp. 51–73.

Mellor, D. H. (2000), 'The Semantics and Ontology of Dispositions', *Mind*, 436, pp. 757–80.

Mill, J. S. (1874), *A System of Logic, Ratiocinative and Inductive*, New York, Harper and Brothers, 8th ed.

Miller, Richard W. (1987), *Fact and Method: Explanation, Confirmation and Reality in the Natural and Social Sciences*, Princeton, Princeton University Press.

Millikan, Ruth (1984), *Language, Thought, and Other Biological Categories: New Foundations for Realism*, Cambridge, MA, MIT Press.

Mumford, Stephen (1998), *Dispositions*, Oxford, Clarendon Press.

Newton, I. (1687), *Philosophiae Naturalis Principia Mathematica*, London.

Newton, I. (1704), *Opticks*, London.

Newton-Smith, W. (1981), *The Rationality of Science*, Boston, Routledge and Kegan Paul.

Noonan, H. (1993), 'Constitution is Identity', *Mind*, 102, p. 133–47.

Nozick, R. (1981), *Philosophical Explanations*, Cambridge, Belknap Press.

Oldroyd, D. (1986), *The Arch of Knowledge. An Introductory Study of the History, Philosophy, and Methodology of Science*, New York, Methuen.

Olson, E. (1997), *The Human Animal: Personal Identity Without Psychology*, Oxford, Oxford University Press.

Papineau, David (1979), *Theory and Meaning*, Oxford, Clarendon Press.

Papineau, David (1996a), 'Theory-Dependent Terms', *Philosophy of Science*, 63, pp. 1–20.

Papineau, David (1996b), 'Philosophy of Science', in *The Blackwell Companion to Philosophy*, N. Bunnin and E. P. Tsui-James (eds), Oxford, Blackwell, pp. 290–324.

Parrini, Paolo (1994a) [1990], 'On Kant's Theory of Knowledge: Truth, Form, Matter', in *Kant and Contemporary Epistemology*, Paolo Parrini (ed.), University of Western Ontario Series in Philosophy of Science, Dordrecht, Kluwer, pp. 195–230.

Parrini, Paolo (1994b), 'With Carnap Beyond Carnap. Metaphysics, Science and the Realism/Instrumentalism Controversy', in W. C. Salmon and G. Wolters (eds) *Logic, Language, and the Structure of Scientific Theories. Proceedings of the Carnap-Reichenbach Centennial, University of Konstanz, 21–4 May 1991*, Pittsburgh/ Konstanz, University of Pittsburgh Press and Universitätsverlag Konstanz, pp. 255–77.

Parrini, Paolo (1998) [1995], *Knowledge and Reality. An Essay in Positive Philosophy* [*Conoscenza e realtà. Saggio di filosofia positiva, Roma-Bari, Laterza*], English translation by Paolo Baracchi, Dordrecht, Kluwer.

Preti, Giulio (1946), 'I limiti del neopositivismo', *Studi Filosofici*, 7/6, pp. 87–96.

Preti, Giulio (1974), 'Lo scetticismo e il problema della conoscenza', *Rivista [critica] di storia della filosofia*, 29, pp 3–31; 123–43; 243–63.

Prior, E. W., Pargetter, R. J. and Jackson, F. C. (1982), 'Three Theses about Dispositions', *American Philosophical Quarterly*, 19, pp. 251–7.

Psillos, Stathis (1999), *Scientific Realism: How Science Tracks Truth*, London/New York, Routledge and Kegan Paul.

Putnam, H. (1975), *Mind, Language and Reality: Philosophical Papers*, vol. 2, Cambridge, Cambridge University Press.

Putnam, H. (1978), *Meaning and the Moral Sciences*, London, Routledge and Kegan Paul.

Putnam, H. (1979), 'Reflections on Goodman's Ways of World-Making', *Journal of Philosophy*, 76, pp. 603–18.

Putnam, H. (1981), *Reason, Truth and History*, Cambridge, Cambridge University Press.

Putnam, H. (1983), *Realism and Reason, Philosophical Papers*, vol. 3, Cambridge, Cambridge University Press.

Putnam, H. (1985), 'A Comparison of Something with Something Else', *New Literary History*, 17, pp. 61–79.

Putnam, H. (1987), *The Many Faces of Realism*, LaSalle, Open Court.

Quine, W. V. O. (1952), *Methods of Logic*, London, Routledge and Kegan Paul.

Quine, W. V. O. (1953), *From a Logical Point of View*, Cambridge, MA, Harvard University Press, 2nd ed., rev., 1st ed., 1953.

Quine, W. V. O. (1970a), *Philosophy of Logic*, Cambridge, MA., Harvard University Press.

Quine, W. V. O. (1970b), 'Existence', in W. Y. Yourgrau, A. D. Breck (eds)., *Physics, Logic and History*, New York, Plenum Press, pp. 89–103.

Quine, W. V. O. (1975), 'On Empirically Equivalent Systems of the World', *Erkenntnis*, 9, pp. 313–28.

Quine, W. V. O. (1980), 'Two Dogmas of Empiricism', in W. V. O. Quine, *From a Logical Point of View*, Cambridge, MA, Harvard University Press, 4th pr., pp. 20–46.

Railton, Peter (1981), 'Probability, Explanation, and Information', *Synthese*, 48, pp. 233–56.

Ramsey, Frank (1925), 'The Foundations of Mathematics', in *Foundations: Essays in Philosophy, Logic, Mathematics and Economics*, D. H. Mellor (ed.), London, Routledge and Kegan Paul, 1978.

Reichenbach, Hans (1956) [1928], *The Direction of Time*, Berkeley and Los Angeles, University of California Press.

Reichenbach, Hans (1958), *The Philosophy of Space and Time*, New York, Dover. Original German ed., 1928.

Reichenbach, Hans (1978) [1933], 'Kant and Natural Science' ['Kant und die Naturwissenschaft'], in H. Reichenbach, *Selected Writings*, vol. 1, M. Reichenbach and R. S. Cohen (eds), Dordrecht, Reidel, pp. 389—404.

Rescher, N. (1973), *Conceptual Idealism*, Oxford, Blackwell.

Rescher, N. (1977), *Methodological Pragmatism*, Oxford, Blackwell.

Rescher, N. (1978), *Peirce's Philosophy of Science. Critical Studies in His Theory of Induction and Scientific Method*, Notre Dame-London, University of Notre Dame Press.

Rescher, N. (1984), *The Limits of Science*, Berkeley/Los Angeles/London, University of California Press.

Rescher, N. (1987), *Scientific Realism: A Critical Reappraisal*, Dordrecht/Boston, Reidel.

Rescher, N. (1990), *A Useful Inheritance: Evolutionary Aspects of the Theory of Knowledge*, Rowman and Littlefield, Savage.

Rescher, N. (1992), *A System of Pragmatic Idealism (Vol. I: Human Knowledge in Idealistic Perspective)*, Princeton, Princeton University Press.

Rey, Abel (1907), *La Théorie de la physique chez les physiciens contemporains*, Paris, Alcan.

Rey, Georges (1998), 'A Naturalistic A Priori', *Philosophical Studies*, 92, pp. 25—43.

Rorty, R. (1979), *Philosophy and the Mirror of Nature*, Princeton, Princeton University Press.

Rorty, R. (1982a), 'The World Well Lost', in R. Rorty, *Consequences of Pragmatism*, University of Minnesota Press, Minneapolis, 1982, pp. 3—18.

Rorty, R. (1982b), *Consequences of Pragmatism (Essays: 1972—1980)*, Minneapolis, University of Minnesota Press.

Rorty, R. (1991), 'Is Natural Science a Natural Kind?', in R. Rorty, *Objectivity, Relativism, and Truth, Philosophical Papers* (vol. 1), Cambridge, Cambridge University Press, 1991, pp. 46—62.

Rorty, R. (1999), *Philosophy and Social Hope*, London, Penguin.

Russell, Bertrand (1912), *The Problems of Philosophy*, London, Oxford Paperbacks, 1967. Original publ. 1912.

Russell, Bertrand (1922), 'The Problem of Infinity Considered Historically', in *Our Knowledge of the External World*, London, George Allen and Unwin, pp. 159—88.

Russell, Bertrand (1948), *Human Knowledge: Its Scope and Limits*, New York, Simon and Schuster.

Russell, Bertrand (1990) [1899], 'The Axioms of Geometry' ['Sur les Axiomes de la Géométrie'], in B. Russell, *Philosophical Papers 1896-99*, N. Griffin and A. C. Lewis (eds), textual apparatus prepared by W. G. Stratton, London, Unwin Hyman, pp. 390—415 [French text on pp. 432—51].

Sainsbury, M. (forthcoming), 'Referring Descriptions'.

Salmon, Wesley C. (1967), 'Carnap's Inductive Logic', *Journal of Philosophy*, 64, pp. 725—39.

Salmon, Wesley C. (1969), 'Partial Entailment as a Basis for Inductive Logic', in *Essays in Honor of Carl G. Hempel*, Nicholas Rescher (ed.), Dordrecht, D. Reidel, pp. 47—82.

Salmon, Wesley C. (1981), 'Causality: Production and Propagation', in *PSA 1980*, P. D. Asquith and R. N. Giere (eds), East Lansing, Michigan, Philosophy of Science Association, pp. 49–69. Reprinted in Salmon, 1998, pp. 285–301.

Salmon, Wesley C. (1994), 'Causality Without Counterfactuals', *Philosophy of Science*, 61, pp. 297–312. Reprinted in Salmon, 1998, pp. 248–60.

Salmon, Wesley C. (1997), 'Causality and Explanation: A Reply to Two Critiques', *Philosophy of Science*, 64, pp. 461–77.

Salmon, Wesley C. (1998), *Causality and Explanation*, New York, Oxford University Press.

Sankey, Howard (1997), *Rationality, Relativism and Incommensurability*, Aldershot, Ashgate Publishing.

Sankey, Howard (2000), 'Methodological Pluralism, Normative Naturalism and the Realist Aim of Science', in Robert Nola and Howard Sankey (eds), *After Popper, Kuhn and Feyerabend: Issues in Recent Theories of Scientific Method*, Dordrecht, Kluwer, pp. 211–29.

Sapir, Edward (1951) 'The Status of Linguistics as a Science' (1929, in E. Sapir, *Selected Writings of Edward Sapir in Language, Culture and Personality* (1949), D. G. Mandelbaum (ed.), Berkeley/Los Angeles/London, University of California Press), pp. 160–66.

Schlick, Moritz (1974) [1918, 1925], *General Theory of Knowledge [Allgemeine Erkenntnislehre]* English translation of the 2nd revised ed. by A. E. Blumberg, with an Introduction by A. E. Blumberg and H. Feigl, Wien/New York, Springer-Verlag.

Schlick, Moritz (1979) [1932, 1938] , 'Form and Content. An Introduction to Philosophical Thinking' (1938), in M. Schlick, *Philosophical Papers*, vol. 2 (1925–36), H. L. Mulder and B. F. B. Van De Velde-Schlick (eds), 'Vienna Circle Collection', Dordrecht, Reidel (Kluwer), 1979, pp. 285–369.

Sellars W. (1963), 'Philosophy and the Scientific Image of Man', in W. Sellars, *Science, Perception and Reality*, London, Routledge and Kegan Paul, 1963, pp. 1–40.

Sellars, W. (1968), *Science and Metaphysics. Variations on Kantian Themes*, London, Routledge and Kegan Paul.

Sellars, W. (1997), *Empiricism and the Philosophy of Mind*, Cambridge, MA, London, Harvard University Press.

Shoemaker, S. (1984), 'Personal Identity: A Materialist's Account', in S. Shoemaker and R. Swinburne, *Personal Identity*, Oxford, Blackwell.

Siegel, Harvey (1990), 'Laudan's Normative Naturalism', *Studies in History and Philosophy of Science*, 21, pp. 295–313.

Snowdon, P. F. (1991), 'Personal Identity and Brain Transplants', in *Human Beings*, D. Cockburn (ed.), Cambridge: Cambridge University Press, 1991.

Sosa, Ernest and Michael Tooley (eds) (1993), *Causation*, Oxford, Oxford University Press.

Stapp H. (1993), *Mind, Matter, and Quantum Mechanics*, Berlin/Heidelberg/New York, Springer-Verlag.

Stich, Stephen P. (1983), *From Folk Psychology to Cognitive Science: The Case Against Belief*, Cambridge, MA, MIT Press.

Stove, David (1991), *The Plato Cult and Other Philosophical Follies*, Oxford, Blackwell.

Strawson, P. F. (1959), *Individuals. An Essay in Descriptive Metaphysics*, London, Methuen.

Suppe, Frederick (1977) [1973], 'The Search for Philosophical Understanding of Scientific Theories' and 'Afterword — 1977', in *The Structure of Scientific Theories*, edited with a Critical Introduction and Afterword by F. Suppe, Urbana/Chicago/London, University of Illinois Press, pp. 1–241, 615–730.

Swinburne, R. (1984), 'Personal Identity: The Dualist Theory', in S. Shoemaker and R. Swinburne, *Personal Identity*, Oxford, Blackwell.

Tarski, Alfred (1994) [1943], 'The Semantic Conception of Truth', in Horwich (1994), pp. 107–41.

Trigg, R. (1989), *Reality at Risk: A Defence of Realism in Philosophy and the Sciences*, New York/London, Harvester Wheatsheaf, 2nd ed.

van Fraassen, Bas C. (1980), *The Scientific Image*, Oxford, Clarendon Press.

van Fraassen, Bas C. (1985), 'Empiricism in the Philosophy of Science', in *Images of Science: Essays on Realism and Empiricism, with a Reply from Bas C. van Fraassen*, Paul M. Churchland and Clifford A. Hooker (eds), Chicago, University of Chicago Press, pp. 245–308.

van Fraassen, Bas C. (1989), *Laws and Symmetry*, Oxford, Clarendon Press.

Van Inwagen, Peter (1990), *Material Beings*, Ithaca, Cornell University Press.

Von Wright, G. H. (1971), *Explanation and Understanding*, Ithaca, NY, Cornell University Press.

Weinberg, S. (1992), *Dreams of a Final Theory*, New York, Pantheon Books.

Whitehead, A. N. and Bertrand Russell (1910–13), *Principia Mathematica*, 3 vols, Cambridge, Cambridge University Press.

Wiggins, D. (1980), *Sameness and Substance*, Cambridge, MA, Harvard University Press.

Williams, B. (1973), 'Are Persons Bodies?', in *Problems of the Self*, Cambridge, Cambridge University Press.

Williams, B. (1986), 'Wittgenstein and Idealism', in *Understanding Wittgenstein*, London, Macmillan, 1974, pp. 76–95.

Williams, Michael (1993), 'Realism and Scepticism', in *Reality, Representation, and Projection*, John Haldane and Crispin Wright (eds), New York, Oxford University Press, pp. 193–214.

Wolff, Robert P. (1963), *Kant's Theory of Mental Activity. A Commentary on the Transcendental Analytic of the 'Critique of Pure Reason'*, Cambridge, MA, Harvard University Press.

Worrall, J. (1989), 'Structural Realism: The Best of Both Worlds?', *Dialectica*, 43/1, pp. 99–125.

Index of Names

Index of Names

Printed and bound by CPI Group (UK) Ltd, Croydon, CR0 4YY

22/10/2024

01777620-0015